MODELLING OF MARINE SYSTEMS

FURTHER TITLES IN THIS SERIES

Elsevier Oceanography Series, 10

MODELLING OF
MARINE SYSTEMS

Edited by

JACQUES C.J. NIHOUL

Mathematical Institute
University of Liège
Liège, Belgium

ELSEVIER SCIENTIFIC PUBLISHING COMPANY
Amsterdam - Oxford - New York 1975

ELSEVIER SCIENTIFIC PUBLISHING COMPANY
335 Jan van Galenstraat
P.O. Box 211, Amsterdam, The Netherlands

AMERICAN ELSEVIER PUBLISHING COMPANY, INC.
52 Vanderbilt Avenue
New York, New York 10017

Library of Congress Card Number: 74-77585

ISBN: 0-444-41232-8

With 74 illustrations and 13 tables

Copyright © 1975 by Elsevier Scientific Publishing Company, Amsterdam

Printed in The Netherlands

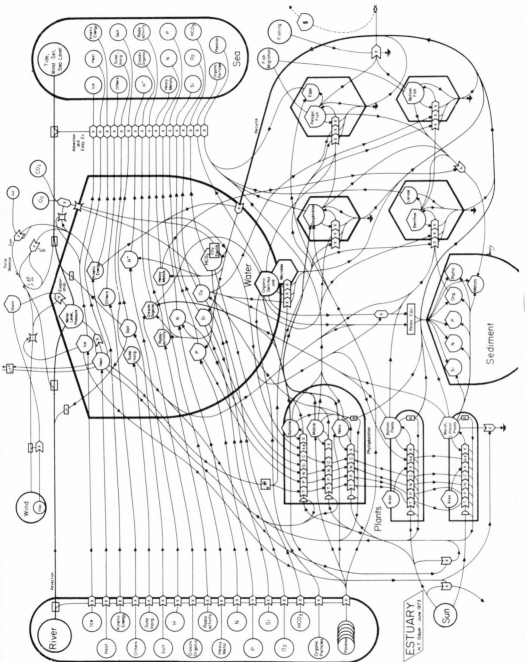

Model for an estuarine system translated into energy circuit language.

PREFACE

The present book emerged from the Conference on Modelling of Marine Systems held in Ofir (Portugal) in June 1973. The conference was organized under the auspices of the NATO Science Committee as part of its continuing effort to promote the useful progress of science through international cooperation.

The Science Committee Conferences are deliberately designed and structured to focus expert attention on what is not known, rather than what is known. The participants are carefully selected to bring together a variety of complementary viewpoints. Through intensive group discussion, they seek to reach agreement on conclusions and recommendations for future research which will be of value to the scientific community.

The attractiveness of this format was confirmed in the present case. Some twenty-eight papers, often specially written reviews, were contributed by the participants for advance circulation, to outline the state-of-the-art in the areas of physical, chemical and biological modelling, and to focus attention on key problems. The availability of this background material precluded the need for lengthy introductory presentations and permitted rapid initiation of interdisciplinary discussions. All participants gave generously and enthusiastically of their expertise and effort during the week of the meeting, and we extend to them our deep gratitude.

We are pleased to have this opportunity to record our special thanks to Prof. Jacques C.J. Nihoul for his diligent efforts as Chairman of the meeting, to his colleagues on the Organizing Committee — Prof. I. Elskens, Prof. D. Garfinkel, Prof. E.D. Goldberg, Dr. R.F. Henry, Prof. J. Pinto Peixoto, Dr. J.H. Steele, Prof. J.W. Stewart — for their wise guidance, and to the leaders and rapporteurs of the Working Groups, as listed, for their indispensable dedication.

We also wish to record our appreciation to the U.S. Office of Naval Research, whose provision of travel support contributed significantly to the success of the meeting.

EUGENE G. KOVACH

Deputy Assistant Secretary General
for Scientific Affairs

FOREWORD

Mathematical models of marine systems have been extensively developed in the recent years. These models were either research models, aiming at a better understanding of the systems' dynamics, or management models designed to assist the administration of water resources and the fight against pollution. At first, the models concentrated on physical, chemical or biological processes according to their particular concerns. Then, the increased threat on the environment requiring a more thorough understanding of ecosystems, the models were extended, in an effort to overlap the frontiers between the disciplines and include imperatives from other fields. Exhaustive multidisciplinary models were conceived which were sometimes praised sometimes criticized for their ambition.

The prodigious development of numerical techniques and computing facilities, recently, supported the idea that such ambitious models were not unrealistic and could provide a convenient framework for the rational assemblage of the so far dispersed partial models.

The time had come for scientists of different fields or different concerns to compare their different approaches to modelling and set up a common language, promoting interdisciplinary research and combined action.

Although rapid progress in modelling of marine systems was evident, it was also obvious that significant developments within any speciality had not always been recognized for their pertinence to the others. The separate evolutions of the different types of models had furthermore concealed many problems which were progressively exposed as the more exhaustive interdisciplinary models stumbled over them.

The desirability of bringing together specialists from all fields of marine modelling became apparent to many within the scientific community. The members of the Organizing Committee were all fully aware of this necessity and they proposed therefore to hold a conference with the intention of assembling a group of active scientists from different countries, to foster a mutually beneficial exchange of information. Such communication, attempting to assort different points of view, was expected to disclose interdisciplinary problems and interdisciplinary solutions unperceived sofar and to identify simultaneously subjects for further research and new paths to further achievement.

To enhance the degree of interaction — after an indispensable plenary introduction to acquaint all participants with the semantics and present status of research within the different fields — small working groups were formed to discuss specialized topics chosen for their ability to provoke a maximum overlap between the different approaches. To prepare and nourish the discussions, all participants were asked to submit in advance a paper (original contribution, review or recently published work) and the selected papers assembled in a pre-Conference volume were made available to everyone before the Conference.

After the Conference, it was recommended that a book be written.

The intention of this book is not the publication of the Conference's proceedings or the reproduction of the informal pre-Conference volume, many papers of which were in the process of being published in specialized journals or were in a preliminary form suitable only to workshop's discussion. The purpose of the book is, beside the diffusion of the conclusions and recommendations of the working group, — providing guide-lines for further theoretical, experimental and applied work — , the survey, through specially commissioned papers, of the state-of-the-art in interdisciplinary modelling of marine systems. The invited papers contributing to this survey are based on the most recent publications and in particular on the content of the pre-Conference volume which had stimulated the discussions. Due reference to the participants' work is given in the text and a detailed list of addresses is included to enable the reader to get directly in touch with any specialist he may wish to consult.

The group reports and the commissionned papers have tried to be accessible to readers of all backgrounds and all disciplines in marine modelling. The intention was more to inform as accurately and clearly as possible specialists of other fields of important developments in one's own domain than to provide fellow scientists with advanced reviews. If mathematical equations have been mentioned in Part I for the sake of assorting the notations between specialists, they are, in most cases, not essential to the general philosophy of the text, which means to be understood without going into the details of them.

The purpose of the present book is thus to assemble the elements of a first manual on modelling of marine systems.

Such a manual could only emerge with the right degree of simplicity and sophistication from such an enthusiastic mixing of active scientists from related but not identical disciplines and concerns.

It is a pleasure, as the Chairman of the Organizing Scientific Committee, to thank all the persons who made that meeting possible; to the Nato Scientific Committee for its generous support, to Dr. Kovach and Miss Austin for their gentle efficiency, to Prof. Peixoto, Prof. Juerra and Dr. Meira for their noteworthy hospitality, to the members of the Organizing Committee for

their dedicated work in preparing and animating the Conference, to all participants finally for their valuable and enthusiastic contributions.

JACQUES C.J. NIHOUL

Chairman of the
Organizing Scientific Committee
December 1973

LIST OF PARTICIPANTS

Mr. Y.A. ADAM
Institut de Mathématique
Université de Liège
Av. des Tilleuls 15
4000 Liège (Belgium)

Dr. N.R. ANDERSEN
Ocean Science and Technology Division
(Code 480)
Office of Naval Research
800 N. Quincy Street
Arlington, Va. 22217 (U.S.A.)

Mr. W. BAYENS
Analytische Scheikunde
Vrije Universiteit Brussel
A. Buyllaan 105
1050 Brussels (Belgium)

Dr. M. BERNHARD
Laboratorio per lo Studio della
Contaminazione Radioattiva del Mare
C.N.E.N. — EURATOM
I—19030 Fiascherino
La Spezia (Italy)

Dr. SANDRA L. BUCKINGHAM
Institute of Resource Ecology
University of British Columbia
Vancouver 8, B.C. (Canada)

Dr. J.D. BURTON
Department of Oceanography
The University
Southampton, SO9 5NH (Great Britain)

Dr. D.H. CUSHING
Fisheries Laboratory
Ministry of Agriculture, Fisheries and Food
Lowestoft, Suffolk (Great Britain)

Prof. Dr. L. DE CONINCK
Rijksuniversiteit Gent
Ledeganckstraat 35
9000 Ghent (Belgium)

Prof. Dr. A. DISTÈCHE
Institut E. Van Beneden
Section: Océanologie
Université de Liège
Quai Van Beneden 22
4000 Liège (Belgium)

Dr. D.M. DI TORO
Environmental Engineering Program
Manhattan College
Bronx, N.Y. 10471 (U.S.A.)

Dr. J. DUGAN
Code 8301
Ocean Sciences Division
Naval Research Laboratory
Washington, D.C. 2P375 (U.S.A.)

Prof. Dr. R. DUGDALE
Dept. of Oceanography
University of Washington
Seattle, Wash. 98105 (U.S.A.)

Prof. Dr. I. ELSKENS
Analytische Scheikunde
Vrije Universiteit Brussel
A. Buyllaan 105
1050 Brussels (Belgium)

Dr. E.J. FEE
Freshwater Institute
501 University Crescent
Winnipeg, Man. (Canada)

Dr. H.G. GADE
Geophysics Institute
University of Bergen
Allegt 70
5000 Bergen (Norway)

Prof. Dr. D. GARFINKEL
Moore School of Electrical Engineering
University of Pennsylvania
Philadelphia, Pa. 19174 (U.S.A.)

Prof. Dr. E.D. GOLDBERG
Scripps Institution of Oceanography
La Jolla, Calif. 92037 (U.S.A.)

Dr. N.S. HEAPS
Institute of Coastal Oceanography and
Tides
Bidston Observatory
Birkenhead, Cheshire L43 7RA
(Great Britain)

Dr. R.F. HENRY
Department of the Environment
Pacific Region
Marine Sciences Directorate
512 Federal Building
Victoria, B.C. (Canada)

Dr. T.S. HOPKINS
Institute of Oceanographic and Fishing
Research (IOKAE)
Aghios Kosmas
Ellinikon (Greece)

Dr. J.L. HYACINTHE
C.N.E.X.O.
29N Plouzane
B.P. 337
Brest (France)

Dr. B-O. JANSSON
Asko Laboratory
c/o Department of Zoology
University of Stockholm
P.O. Box 6801
11386 Stockholm (Sweden)

Dr. B.H. KETCHUM
Associate Director
Woods Hole Oceanographic Institution
Woods Hole, Mass. 02543 (U.S.A.)

Prof. Dr. W. KRAUSS
Institut für Meereskunde an der
Universität Kiel
Düsternbrooker Weg 20
23 Kiel (Germany)

Dr. P. LEBLOND
Institute of Oceanology
University of British Columbia
Vancouver 8, B.C. (Canada)

Dr. J.J. LEENDERTSE
RAND Corporation
1700 Main Street
Santa Monica, Calif. 90406 (U.S.A.)

Dr. P.S. LISS
School of Environmental Sciences
University of East Anglia
Norwich, NOR 88C (Great Britain)

Dr. I.N. McCAVE
School of Environmental Sciences
University of East Anglia
Norwich, NOR 88C (Great Britain)

Dr. J.C. MACKINNON
Department of Biology
Dalhousie University
Halifax, N.S. (Canada)

Dr. K.H. MANN
Department of Biology
Dalhousie University
Halifax, N.S. (Canada)

Prof. Dr. R. MARGALEF
Instituto de Investigaciones Pesqueras
Paseo Nacional s/n
Barcelona 3 (Spain)

Dr. D.W. MENZEL
Skidaway Institute of Oceanography
P.O. Box 13687
Savannah, Ga. 31406 (U.S.A.)

Mr. J-P. MOMMAERTS
Dienst Ekologie en Systematiek
Vrije Universiteit Brussel
A. Buyllaan 105
1050 Brussels (Belgium)

Prof. Dr. C. MORELLI
Presidente
Osservatorio Sperimentale di Geofisica
34123 Trieste (Italy)

Dr. C.H. MORTIMER
Center for Great Lakes Studies
University of Wisconsin—Milwaukee
Milwaukee, Wisc. 53201 (U.S.A.)

Dr. K. MOUNTFORD
Benedict Estuarine Laboratory
Academy of Natural Sciences of
Philadelphia
Benedict, Md. 20612 (U.S.A.)

Dr. T.S. MURTY
Marine Sciences Directorate
615 Booth Street
Ottawa, Ont. (Canada)

Prof. Dr. J.C.J. NIHOUL, Chairman
Belgian National Program on the
Environment
Ministry for Science Policy, Belgium
Institut de Mathématique
Université de Liège
Av. des Tilleuls 15
4000 Liège (Belgium)

Prof. Dr. J.J. O'BRIEN
Department of Meteorology
Florida State University
Tallahassee, Fla. 32306 (U.S.A.)

Prof. Dr. H.T. ODUM
Department of Environmental Engineering
University of Florida
Gainesville, Fla. 32601 (U.S.A.)

Mr. J.P. O'KANE
Department of Civil Engineering
University College
Upper Merrion Street
Dublin 2 (Ireland)

Dr. R.T. PAINE
Department of Zoology
University of Washington
Seattle, Wash. 98105 (U.S.A.)

Prof. Dr. T.R. PARSONS
Institute of Oceanography
University of British Columbia
Vancouver, B.C. (Canada)

Mr. G. PICHOT
Institut de Mathématique
Université de Liège
Av. des Tilleuls 15
4000 Liège (Belgium)

Prof. Dr. J. PINTO PEIXOTO
Institute of Geophysics
University of Lisbon
Lisbon (Portugal)

Dr. A. PIRO
Laboratorio per lo Studio della
Contaminazione Radioattiva del Mare
C.N.E.N. — EURATOM
I—19030 Fiascherino
La Spezia (Italy)

Dr. T. PLATT
Marine Ecology Laboratory
Bedford Institute of Oceanography
Dartmouth, N.S. (Canada)

Prof. Dr. Ph. POLK
Fakulteit der Wetenschappen
Dienst Ekologie en Systematiek
Vrije Universiteit Brussel
A. Buyllaan 105
1050 Brussels (Belgium)

Dr. H. POSTMA
Netherlands Institute for Sea Research
Postbus 59
Horntje, Texel (The Netherlands)

Dr. G. RADACH
Sonderforschungsbereich 94
Meeresforschung Hamburg
Universität Hamburg
Heimhuderstrasse 71
2 Hamburg 13 (Germany)

Dr. E.D. SCHNEIDER
National Marine Water Quality Laboratory
U.S. Environmental Protection Agency
P.O. Box 277
West Kingston, R.I. 02892 (U.S.A.)

Dr. J.H. STEELE
Marine Laboratory
Department of Agriculture and Fisheries
for Scotland
P.O. Box 101
Aberdeen, AB9 8DB (Great Britain)

Prof. Dr. R.W. STEWART
Marine Sciences Branch
Pacific Region
Department of the Environment
1230 Government Street
Victoria, B.C. (Canada)

Dr. R.E. ULANOWICZ
University of Maryland
Natural Resources Institute
Chesapeake Biological Laboratory
Box 36
Solomons, Md. 20688 (U.S.A.)

Dr. J.J. WALSH
Department of Oceanography
University of Washington
Seattle, Wash. 98195 (U.S.A.)

Dr. C.J. WALTERS
Institute of Resource Ecology
University of British Columbia
Vancouver 8, B.C. (Canada)

Prof. Dr. R. WOLLAST
Laboratoire de Chimie Industrielle
Université Libre de Bruxelles
Avenue F.D. Roosevelt 50
1050 Brussels (Belgium)

Dr. B. ZEITZSCHEL
Institut für Meereskunde an der
Universität Kiel
Düsternbrooker Weg 20
23 Kiel (Germany)

Observers

Cdt. M. RENSON
Services du Premier Ministre
Commission Interministérielle de la
Politique Scientifique
Environnement
Rue de la Science 8
1040 Brussels (Belgium)

Dr. W. ZAHEL
Institut für Meereskunde
Heimhudstrasse 71
2 Hamburg 13 (Germany)

Dr. VICTOR TAVARES
Serviço Meteorologico Nacional
Rua Saraiva Carvalho No. 2
Lisbon (Portugal)

Dr. ARTUR PIRES
Serviço Meteorologico Nacional
Rua Saraiva Carvalho No. 2
Lisbon (Portugal)

Dra. ISABEL AMBAR
Instituto Geofisico do Infante D. Luis
Faculdade de Ciencias de Lisboa
Rua da Escola Politecnica
Lisbon (Portugal)

Dr. ARMANDO FIUZA
Instituto Geofisico do Infante D. Luis
Faculdade de Ciencias de Lisboa
Rua da Escola Politecnica
Lisbon (Portugal)

Dr. DANIEL RODRIGUES
Instituto Hidrografico
Ministerio da Marinha
Rua das Trinas
Lisbon (Portugal)

Cte. JOSÉ MANUEL SALDANHA
Instituto Hidrografico
Ministerio da Marinha
Rua das Trinas
Lisbon (Portugal)

Prof. Dr. MARIA HELENA GALHANO
Instituto de Zoologia Dr. Augusto Nobre
Faculdade de Ciencias do Porto
Porto (Portugal)

Dr. LUIS SALDANHA
Museu Bocage — Faculdade de Ciencias
Rua da Escola Politecnica
Lisbon (Portugal)

Dr. JOAQUIM AGUAS
Departamento de Higiene
Escola Superior de Medicina Veterinaria
Rua de Gomes Freire
Lisbon (Portugal)

Prof. Dr. ARNALDO ROZEIRA
Faculdade de Ciencias do Porto
Porto (Portugal)

Prof. Dr. JORGE VEIGA
Laboratorio de Quimica
Universidade de Lourenço-Marques
Lourenço-Marques (Moçambique)

Dr. CARLOS PISSARRO
Centro de Biologia Aquatica Tropical
Junta de Investigaçoes do Ultramar
Rua Dr. Antonio Candido, 9
Lisbon (Portugal)

NATO

Dr. E.G. KOVACH
Deputy Assistant Secretary
General for Scientific Affairs,
NATO
1110 Brussels (Belgium)

Miss E.I. AUSTIN
Scientific Affairs Division
NATO
1110 Brussels (Belgium)

CONTENTS

Part III. Reports of the working groups and recommendations for future work

PART I

CONCEPTS AND TECHNIQUES OF MARINE MODELLING

CHAPTER 1

MARINE SYSTEMS ANALYSIS

Jacques C.J. Nihoul

1.1. INTRODUCTION

The presence of many strongly interdependent variables makes ecosystems difficult to describe. The simple collection of data and their descriptive ordering are such formidable tasks that one has often ignored the need for doing more than this.

The present alarming state of ecological problems, however, calls for a more thorough understanding and a more rational and strict control of the environment.

Monitoring, control and management are achievable only if, for a selected number of representative variables, one can predict evolution and, within appropriate constraints and tolerances, one can find, by optimization, the necessary compromises between the requirements of increasing industrialization and affluent society and the necessity to preserve the valuable natural resources.

To predict the evolution of the selected variables, one must have some modelled idea of their behaviour.

A model can be of many sorts. It can be a fairly *literal model* like a mechanical model demonstrating the possible motions of an animal's limb; it can be a *scale model* like an instrumented reduced-scale aircraft in a wind tunnel; it can be a more elaborate *physical model* operating in quite another energy sphere as an electrical network simulating dynamical interactions in the food chain. Such models may be called *iconic models*.

The model can also be a *mathematical model.* The mathematical model is the ultimate goal. Indeed, if several distinct physical processes can be used to simulate the same phenomenon, it is because, fundamentally, they are described by the same equations and that the same equations (submitted to the same boundary and initial conditions) yield the same solution independently of the particular significance of the variable, whether it represents an electrical potential, a stream function, a temperature, etc.

Hence one mathematical model incorporates all the equivalent physical

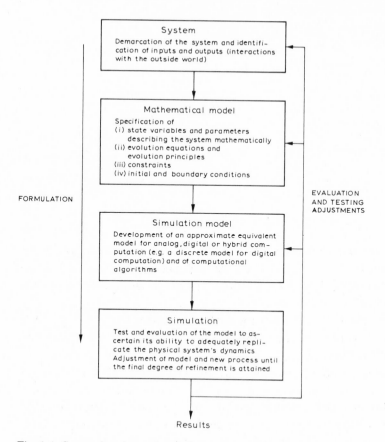

Fig. 1.1. Successive steps of a mathematical simulation process.

models. It is appropriate for analog and digital computation and plays a key role in the general process of *mathematical simulation.*

The successive steps of a mathematical simulation process are briefly described in Fig. 1.1. In the next sections, each of them will be examined in connection with the simulation of the marine system.

1.2. DEMARCATION OF THE MARINE SYSTEM

To model a marine system, it is first required to define the system without ambiguity, separate it from the "outside world" and identify the exchanges between the system and the exterior (inputs and outputs).

The definition of the system implies, in the first place, the specification of its geographical configuration and of its life time, i.e., the system must be clearly situated in the physical $x - t$ space. Clearly however this is not

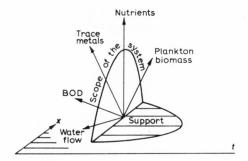

Fig. 1.2. Support and scope of a marine system.

enough to define the system for the purpose of modelling. It merely says where the system is, not what it is.

What the system is, in reality, is obviously terribly complex and this is the reason for making a model. The model must first identify which properties of the real system are essential and how many of them are required to give, in agreement with the model's objective, a satisfactory (the simplest satisfactory) description of the system. In other words, the model must identify the system for its own purpose and specify what variables are considered necessary and sufficient to describe, with the required accuracy, the *state of the system*.

The demarcation of the system implies thus the delimitation of its position in $x - t$ space (the *support* of the system) and the specification of its span in the state of possible state variables (the *scope* of the system). The situation is summarized sketchily in Fig. 1.2.

Support of the system

Ideally, the marine system, limited at the air—sea interface and at the sea floor, should cover the entire ocean for all time. Clearly, this is not reasonable and, in practice, one is interested in modelling, during some given period of time, an estuary, a gulf, a coastal region, a channel, a sea, according to one's particular design. There is thus a natural geographical and time limitation of the marine system.

Scope of the system

The scope of the system is directly related to its ambition. Simple estuarine models may be content with BOD (Biological Oxygen Demand) as sole state variable. Other models will be more sophisticated and include a greater variety of chemical and ecological variables.

It is conceivable to divide the marine system into three sub-systems, physical, chemical and biological, with the necessary input—output links between them. Most models in the past have more or less conformed to this simplified view. One realizes now — in particular when faced with pollution problems — the limitations of such models and the necessity of a truly interdisciplinary approach.

Nevertheless, inasmuch as it is confined in physical space, the system must be limited in scope to the state variables which are essential to describe its behaviour in relation with one's particular concern.

The same *natural* system may thus be represented by several *model* systems which will have normally the same support but which will differ by their scope and which will generate different types of mathematical models, accorded to their particular designs. These models must all be regarded as "sub-sets" of some — still tentative — general model; the construction of which must be pursued to keep track of all the aspects which have been sacrificed to urge conclusive, though partial, results.

Reduction of the size of the system

In many cases, one is not interested in, or equipped for, the detailed dynamics of the system but only in its average or global behaviour in some sense.

Averaging can be performed in state space as in physical space. Many estuarine models, for instance, restrict attention to the time and downstream evolution of depth- (and width-) averaged properties. This amounts to reducing the support to one space and one time dimension. In a first stage, a model may be limited to the description of salinity and turbidity which may be regarded as the total concentrations of substances in solution and in suspension, respectively. In a later stage, the model may be perfected by including the global concentrations of essential chemicals (nutrients, pollutants, etc.) in solution and in suspension or by extending the analysis to the food chain, considered as a whole or as a limited number of groups of living species (phytoplankton, zooplankton, carnivores, etc.). In these cases, a reduction is effected in the scope of the system.

A region of space which is described only by its average or global properties is called a "box" in physical space and a "compartment" in state space. The transfer of one property from one box to another is called a "flow", the transfer from one compartment to another, a "translocation".

In the examples given above, each cross-section of the estuary is treated as a box, while dissolved substances, suspensions, divisions of the food chain are treated as compartments.

It will be shown that the global description of any physical region as a box, characterized by its sole average properties, can only be of value if the system is reasonably uniform inside the box. It is thus often necessary to subdivide the system's support in many boxes or "niches" where similar environmental conditions prevail. This process amounts in fact to reducing the dimensions of the system in physical $x - t$ space while increasing its dimensions in state space. For instance, a depth—width-averaged model of an estuary may be further simplified by assuming fairly homogeneous conditions in each of r "reaches" along the direction of the stream. The characteristics of each reach are now solely dependent on time (and the support is reduced to some interval of time), but if there are n state variables for each reach, the description of the total system now involves rn state variables.

Frequently, the assumption is made that the system is steady. Steady-state models are usually satisfactory when one is primarily interested in the average properties of the system over some sufficiently long period of time. The assumption of stationarity reduces the support to a region in x space.

Thus, according to the particular design of the model, the demarcation of the system can be more or less severe. It involves in general:

(1) The definition of the support; i.e., the delimitation of a geographical region and of a period of time over which the study will be conducted.

(2) The definition of the scope; i.e., the specification of the state variables which are essential for the particular problem under investigation.

(3) The reduction of the support; i.e., the averaging over one or several space coordinates or the time.

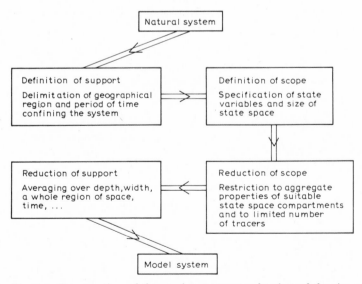

Fig. 1.3. Demarcation of the marine system: reduction of the size of the system.

(4) The reduction of the scope; i.e., the restriction to aggregate properties of suitable compartments in state space.

In addition to average compartment values of the most important state variables, one may wish to follow as such some particular chemical or species which has been found to behave as an illustrative tracer of the evolution of the whole system. Additional state variables may thus be introduced to describe representative tracers. The situation is summarized in Fig. 1.3.

Inputs—outputs

The interactions of the system with the "outside world" provide the *inputs* and *outputs* to the system. The identification of all inputs and outputs is essential to the construction of the mathematical model.

The geographical delimitation of the system introduces coastal and open boundaries where exchanges with the exterior will occur in the form of *boundary inputs* and *outputs*. Boundary interactions also occur at the sea surface and at the sea floor.

In addition to boundary interactions, the outside world influences the system through *volume* forces, sources and sinks (gravitation and tidal forces, solar radiation, dumping of solid wastes or pollutants, fishing, etc.).

The situation is summarized in Fig. 1.4 (where the time dependence is implicit).

The limitation of the scope of the system introduces another form of separation between the system and the outside, and corresponding exchanges also appear as inputs and outputs. This is worth emphasizing because one is naturally tempted to associate the concepts of "interior" and "exterior" to

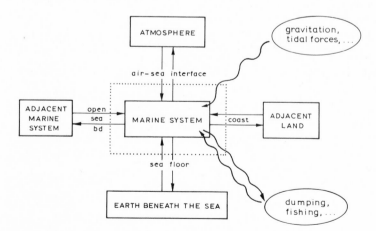

Fig. 1.4. Schematic representation of inputs and outputs.

the physical demarcation of the system only and, erroneously, to regard as internal parts of the system any variable or mechanism which is located in space and time inside the system's support.

In modelling, however, a system is defined by the span of its state variables as much as by its $x - t$ support, and all elements which do not belong to the state space must be treated as part of the outside world even though they are concomitant with the system in space and time.

For instance, as discussed by Steele in Chapter 10, models of primary productivity in some well-delimited region of the sea usually describe the interactions between the total concentration of nutrients in that region and the total biomasses of phytoplankton and zooplankton. Some models (e.g. Pichot, 1973) also include marine bacteria in the state variables, others do not. The bacteria which live in the marine region under investigation may be part of the "natural" system but they belong to the mathematical system only in the first type of models. In the second type, they are definitely part of the outside world and their eventual influence on the system's dynamics must appear as an input from the exterior.

As an example, summarizing the different steps of the system's demarcation, it is illustrative to discuss briefly the paper presented by Adam (1973) and reviewed by Goldberg in this volume (Chapter 8).

Adam is concerned with the evolution of nutrients in shallow seas. In a first attempt to model the mechanism of this evolution, he considers a simplified system defined as follows:

(1) Support: the Ostend "Bassin de Chasse", a closed isolated sea basin of constant water level; a period of time of six months.

(2) Scope: the different forms of phosphorus.

(3) Reduced support: the time period. The basin is treated as a box and only total (or space-averaged) concentrations are considered. Their evolution in time is governed by ordinary differential equations.

(4) Reduced scope: six variables representing the total amounts of phosphorus in six compartments (dissolved phosphates in "free" water, non-living matter in suspension, dissolved phosphates in interstitial water, bottom sediments, plankton, benthos).

The time evolution of the six state (compartment) variables is simulated on a digital computer in different situations, providing new insights on the effects of non-linear interactions (see also Chapter 5 by Heaps and Adam in this volume).

1.3. STATE VARIABLES AND CONTROL PARAMETERS

The first step in constructing a mathematical model is the definition of the support and the scope of the system. The latter implies the selection of a *limited number of representative state variables*. There must be sufficiently few of them for their evolution equations to be amenable to analysis but also enough of them to describe adequately the system's behaviour.

Ideally, if one were willing and able to describe the natural marine system in its enormous complexity, one would have to determine the concentrations and velocities of all forms of chemical substances (dissolved, particulate, etc.) and of all living species at every point of the support.

Such a description is of course unpractical and the first concern of a model, as noted before, is to replace the natural complex system by a simplified one (e.g. Nihoul, 1972).

(1) The first simplification is obtained by introducing bulk variables to describe the mechanics of motion. Sea water being a mixture of dissolved chemicals, particulate matter, living species, if ρ_i and v_i denote respectively the specific mass (mass per unit volume) and the velocity of constituent C_i, one defines:

$$\rho = \Sigma \rho_i \tag{1.1}$$

$$v = \frac{\Sigma \rho_i \, v_i}{\Sigma \rho_i} \tag{1.2}$$

where the sums are over all constituents.

The sea dynamics is analyzed in terms of ρ and v and associated variables like the pressure and the temperature.

(2) The second type of simplification is related to the restriction of chemical and biological variables to those which are essential. In other words, one no longer considers all constituents C_i but a limited number of them.

(3) The scope of the system is then further reduced by renouncing the study of all forms, combinations and varieties of the selected constituents and restricting attention for most of them to their aggregate concentrations in a certain number of compartments; a very limited number of them being eventually still studied in details as tracers.

In addition, the total or specific masses of the various compartments must be known. Particularly important in this respect are the specific mass of all dissolved substances (the *salinity*) ρ_s, the specific mass of the suspensions (the *turbidity*) ρ_t, the specific biomasses of the living compartments like phytoplankton, zooplankton, etc.; the distinction between salinity and turbidity being, in a sense, arbitrary as the experimentalist will call dissolved everything which goes through a filter of pre-decided fineness.

The state variables introduced here are noted ρ_α. For convenience, ρ_α will be referred to as the specific mass of "constituent" α; the word constituent being used in an extended sense as it may denote a whole aggregate (all dissolved substances, etc.) or the aggregate content of a compartment in a specific chemical (nutrient, pollutant, etc.).

(4) The influence of temperature on the dynamics of the marine system has already been mentioned. This is a thermodynamic variable deep-rooted in the statistical mechanics of the system.

Ecological variables of similar statistical origin, such as the index of diversity or the index of stability are often useful to characterize the "health" of an ecosystem and yield information which the simple consideration of biomasses and pollutant concentrations is unable to provide. The label "thermodynamic" — inspired from the recent progresses of irreversible thermodynamics in biology and ecology — is often used generically to indicate this type of variable.

(5) The reduction of the support by averaging over one or several space coordinates is responsible for an additional simplification. The average over one spatial direction disposes of the corresponding component of the velocity vector which remains only to be known at the boundaries to estimate the flows in and out of the box. Thus, in depth-averaged models, the velocity vector is reduced to a two-component horizontal vector; in depth- and width-averaged estuarine models, it is reduced to a single component along the main river flow; in box models averaged over a whole space region, the water displacements inside the box become irrelevant, and, provided the inputs and outputs at the frontiers of the box can be estimated, the velocity disappears from the state variables.

The different types of state variables are shown in Fig. 1.5.

A final simplification is introduced by separating each state variable into a mean part (in some appropriate sense defined below) and a fluctuating part

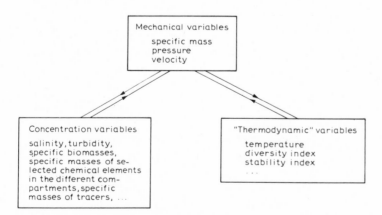

Fig. 1.5. State variables of the marine system.

of which only the general effect — not the details — will appear in the model.

Indeed, one builds a model with the purpose of simulating a specific phenomenon. Let T be its characteristic time scale. Oscillating or erratic processes with characteristic times much smaller than T will tend to cancel each other over a time of order T. Thus they will contribute to the dynamics only through non-linear terms which one may hope to take into account by some appropriate empirical closure ("eddy" diffusivities, etc.). This qualitative argument can be enforced mathematically by introducing a time average — in the sense of the Krylov-Bogoliubov-Mitropolsky method — over some intermediate time period sufficiently small that the phenomenon under investigation can "pass through untouched" but sufficiently large to eliminate the details of the rapid erratic and oscillating processes encumbering the analysis (e.g. Nihoul, 1972).

The word "mean" will be used in that sense in the following, denoting the smooth-running functions obtained by filtering out the fluctuations. Mean quantities will be denoted by angular brackets ⟨ ⟩ or by special symbols. The notations are defined in Table 1.1. It will be assumed that the general effect of the fluctuations can be expressed in terms of the "mean" quantities, eventually with the help of empirical formulas injected into the model.

In addition to the state variables, different kinds of parameters appear inevitably in the mathematical description of the system. These may be called *control parameters* as they influence the evolution of the system (hence appear in the evolution equations) but are not predicted by the model itself (no specific evolution equation is written for them).

The first kind of control parameters one thinks of are the *guidance parameters* which are at the disposal of man to manage the marine system

TABLE 1.1

Symbols for the state variables and their "mean" values

State variable	Symbol	Smooth-running part
velocity	v	$u = \langle v \rangle$
density (specific mass of "mixture")	ρ	$r = \langle \rho \rangle$
pressure	π	$p = \langle \pi \rangle$
temperature	ϑ	$\theta = \langle \vartheta \rangle$
concentration (specific mass of "constituent" α)	ρ_α	$r_\alpha = \langle \rho_\alpha \rangle$
salinity (specific mass of dissolved substances)	ρ_s	$r_s = \langle \rho_s \rangle$
turbidity (specific mass of suspensions)	ρ_t	$r_t = \langle \rho_t \rangle$
specific biomass of		
phytoplankton	ρ_p	$r_p = \langle \rho_p \rangle$
zooplankton	ρ_z	$r_z = \langle \rho_z \rangle$
benthos	ρ_b	$r_b = \langle \rho_b \rangle$
carnivores	ρ_c	$r_c = \langle \rho_c \rangle$
etc.		

according to some optimal design. The guidance parameters will be discussed later in connection with the evolution principles which may in some cases govern, in part, the evolution of the system.

Most of the parameters, however, which control the evolution of the system cannot be chosen to conform with man's concern. They are imposed by nature. These parameters arise from the initial demarcation of the system, the necessity of restricting the state variables and formulating the laws of their evolution in a simple and tractable way. They reflect all the aspects of the natural system of which the model does not take charge; usually because the additional equations required for their prediction would jeopardize the simulation by their difficulty, their dubiousness or simply by increasing the size of the system beyond the computer's ability.

Although they are rarely known beforehand and, in most cases, must be determined approximately by separate models, experimental data or sideways theoretical reflection, the control parameters which result from the closure of the system must be regarded, in the language of the theory of control, as fixed and distinct, therefore, from the guidance parameters mentioned above.

The separation between state variables and control parameters is, of course, more or less arbitrary and a function of the model's capability and ambition.

For instance, all models of primary productivity (state variables: nutrients and plankton) are controlled by the incident light. In a first stage, the incident light may be taken as a fixed control parameter and be given an empirical value. The model can be refined to give the incident light at every depth as a function of the intensity of light at the sea surface, using the transparency of water as a new control parameter. In an even more perfect version of the model, the transparency of water can be included in the state variables and inferred from the turbidity which itself can be predicted by the model.

In his early work on the productivity of the North Sea, Steele (1958) considered a simple model of the stratified waters consisting of two superposed layers of uniform but different densities. The exchanges between the two layers were accounted for by a coefficient of mixing m across the thermocline. The coefficient m is a control parameter. Fig. 1.6 (reproduced from Steele, 1958) shows the nature of the control.

One notes that more mixing decreases the maximum productivity and increases the time necessary to reach the peak value; it also increases the "steady state" production*.

---

* The tendency to approach a summer steady state may however be, according to Steele, an artifact of the mathematics.

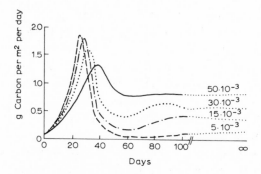

Fig. 1.6. Simulated seasonal cycle of phytoplankton for different values of the mixing rate ($5 \cdot 10^{-3}$, $15 \cdot 10^{-3}$, $30 \cdot 10^{-3}$, $50 \cdot 10^{-3}$). Reproduced from Steele (1958, fig. 6), with the permission of the controller of HM Stationery Office.

The chemical reaction rates may be regarded as control parameters hopefully determined by chemical kinetics, i.e. by laboratory experiments or by some fundamental molecular theory conducted in parallel with the model but not part of it.

The dynamics of translocations (transfer of a chemical element from one compartment to another) must be given appropriate mathematical form. This cannot be done in general without introducing several control parameters, the values of which can only be ascertained experimentally.

In Adam's model for the evolution of phosphorus in a closed sea basin (Adam, 1973), the marine bacteria are not considered among the state variables although they play an important role in dissolving particulate matter. Their influence is accounted for by a control parameter.

As mentioned before, the state variables are separated into mean or slowly varying quantities (as defined above) and fluctuations related to rapid oscillating or erratic motions. The model, set up to describe the mean variables, is strongly affected by the fluctuations which unite in the non-linear terms, even though they may roughly cancel over any period of time characteristic of the mean process.

Rapid oscillations and erratic motions are found to create an enhanced dispersion, similar to the molecular diffusion but many times more effective. In the model of molecular diffusion, coefficients of dispersion ("eddy" diffusivities) are introduced to describe the mixing effect produced by the macroscopic agitation of the water. These coefficients are among the most important control parameters of the model. They are usually given semi-empirical forms partly induced from the observations, partly suggested by the theoretical studies of turbulence.

1.4. EVOLUTION EQUATIONS

The state variables are governed by a system of evolution equations which may be algebraic or differential. They express the complete local budget of mass, momentum, energy, species concentrations, etc.

The diagram shown in Fig. 1.7 recapitulates that the state variables may change in time or in space as a result of releases (from external sources) or withdrawal (to external sinks), internal interactions and displacement of material induced by sea motion or migration.

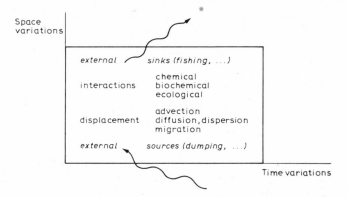

Fig. 1.7. Schematic diagram of the mechanisms responsible for the change in time and space of the state variables.

The term *migration* refers to a motion relative to the bulk motion indicated by the bulk velocity v. In general indeed, the individual velocities v_α are not exactly equal to v. Thus:

$$\rho_\alpha v_\alpha = \rho_\alpha v + \rho_\alpha (v_\alpha - v) \tag{1.3}$$

The first term in the right-hand side of eq. 1.3 is the *flow* of constituent C_α while the second term is the *flux*. The flux is due, in the first place, to the *molecular diffusion*. Molecular diffusion, however, is usually negligible as compared with *turbulent dispersion* which is due to that part of the flow which corresponds to fluctuating erratic motions and which appears explicitly in the evolution equations once the velocity field has been separated into a *mean advecting* part and a fluctuation around the mean. In practice, the molecular diffusion terms are de facto included in the turbulent dispersion terms.

Other contributions to the flux come from the displacements of animals "of their own free will" through the water (horizontal migration of fish, vertical motion of plankton with light, etc.), from the sedimentation of heavy particles or the rising of light constituents or gases.

These types of fluxes are summarized under the general appellation "migration".

Some kinds of migrations (like the sedimentation) can be formulated fairly easily, others (like the migration of fish) are more difficult to express, and often only rudimentary statistical data are available to estimate them.

The evolution equations may be summarized as follows (e.g. Nihoul, 1972):

(1) It is generally admitted that the specific mass ρ obeys a conservation equation of the form:

$$\frac{\partial \rho}{\partial t} + \nabla \cdot \rho \boldsymbol{v} = 0 \tag{1.4}$$

where ∇ is the vector-operator:

$$\nabla \equiv \boldsymbol{e}_1 \frac{\partial}{\partial x_1} + \boldsymbol{e}_2 \frac{\partial}{\partial x_2} + \boldsymbol{e}_3 \frac{\partial}{\partial x_3} \tag{1.5}$$

Eq. 1.4 can be considerably simplified if ρ may be regarded as the sum of a constant reference density ρ_m and a small deviation ρ' from ρ_m due to temperature, salinity, turbidity variations, etc., i.e.:

$$\rho = \rho_m + \rho' \; ; \qquad \rho' \ll \rho_m \tag{1.6}$$

Eq. 1.4 then implies:

$$\nabla \cdot \boldsymbol{v} \sim O\left(\frac{v}{\ell}\frac{\rho'}{\rho_m}\right) \ll \frac{v}{\ell}$$

where v and ℓ are typical velocity and length scales. Hence, for any q:

$$\boldsymbol{v} \cdot \nabla q \gg q \nabla \cdot \boldsymbol{v}$$

This is the basis of the so-called Boussinesq approximation and the origin of the much favoured *incompressibility condition*:

---

* Note that eq.1.8 is valid provided the same q appears on the two sides of the inequality. For instance, $\nabla \cdot v_j \boldsymbol{v} \sim \boldsymbol{v} \cdot \nabla v_j$; $\nabla \cdot \rho_s \boldsymbol{v} \sim \boldsymbol{v} \cdot \nabla \rho_s$; etc. It would be wrong to substitute eq.1.9 in eq.1.4 and write:

$$\frac{\partial \rho'}{\partial t} + \boldsymbol{v} \cdot \nabla \rho' = 0$$

for there is *definitely* a term $\rho_m \nabla \cdot \boldsymbol{v}$ which, owing to the force of circumstances, is of the same order. The equation for ρ' must be derived from the equations of the ρ_α's on which it depends.

$$\nabla \cdot v = 0 \tag{1.7}$$

(2) The evolution equation for the velocity field is obtained from the conservation of momentum and reads:

$$\frac{\partial}{\partial t}(\rho v) + 2\rho \,\Omega \wedge v + \nabla \cdot \rho vv = -\nabla \pi + \rho F + \mu \nabla^2 v \tag{1.8}$$

where $2\,\Omega \wedge v$ is the Coriolis acceleration, π the pressure as before, F the sum of the external forces per unit mass, and μ the molecular viscosity.

If, as it is often the case, eq. 1.6 holds, it is permissible to replace ρ by ρ_m in all the terms of the left-hand side although it is generally necessary to keep ρ in the forcing term ρF. Indeed, F includes the acceleration of gravity g which is quite large compared to typical flow accelerations. Only in perfectly well-mixed seas like perhaps the Southern Bight of the North Sea can ρ be approximated by ρ_m in the gravity term ρg.

The approximation of ρ by ρ_m in all terms of the momentum equation except the forcing term is part of the *Boussinesq approximation*.

Let now the velocity field v be separated into a mean part u and a fluctuating part of zero mean w (cf. section 1.3):

$$v = u + w \tag{1.9}$$

Taking the mean of eq. 1.8 and making the Boussinesq approximation, one gets:

$$\frac{\partial u}{\partial t} + 2\Omega \wedge u + \nabla \cdot uu = -\nabla \left(\frac{p}{\rho_m}\right) + \left\langle \frac{\rho}{\rho_m} F \right\rangle + \mathcal{D} \tag{1.10}$$

introducing:

$$\mathcal{D} = -\nabla \cdot \langle ww \rangle + \frac{\mu}{\rho_m} \nabla^2 u \tag{1.11}$$

where $\langle ww \rangle$ denotes the mean of the quadratic term ww and represents the non-linear effect of the fluctuations, responsible for a dispersion of momentum similar to the molecular diffusion but many times more efficient.

In view of their similar effect, it seems reasonable to express $\mathcal{D}$ on the model of the molecular diffusion term. In a first stage, this is done by introducing a new coefficient of dispersion (e.g., ν), assumed constant, such that:

$$\mathcal{D} = \nu \nabla^2 u \tag{1.12}$$

where ν plays the role of a control parameter. In a more elaborate model, different coefficients of dispersion may be postulated in different directions and they may be considered more generally as functions of position, in which case they cannot be factorized out of the divergence.

Thus, for instance, one shall write ($\mathcal{D}_j$ being any component of $\mathcal{D}$):

$$\mathcal{D}_j = \frac{\partial}{\partial x_1}\left(\nu_1 \frac{\partial u_j}{\partial x_1}\right) + \frac{\partial}{\partial x_2}\left(\nu_2 \frac{\partial u_j}{\partial x_2}\right) + \frac{\partial}{\partial x_3}\left(\nu_3 \frac{\partial u_j}{\partial x_3}\right) \tag{1.13}$$

or even more sophisticated expressions involving more control parameters ν_i.

(3) The concentration variables ρ_α are governed by conservation equations of the form:

$$\frac{\partial \rho_\alpha}{\partial t} + \nabla \cdot \rho_\alpha \, v_\alpha = R_\alpha + J_\alpha \tag{1.14}$$

where R_α and J_α denote the rates of local production (or destruction) of "constituent" α by exterior and interior agents respectively. R_α represents direct exchanges between the outside world and internal parts of the system. It should not be confused with the boundary inputs and outputs which from a mathematical point of view will arise only later to determine the constants of integration in the general solution of eq. 1.14. J_α represents the chemical, biochemical and ecological interactions between "constituents". It creates a coupling between the state variables. The second term in the left-hand side can be written:

$$\nabla \cdot \rho_\alpha v_\alpha = \nabla \cdot \rho_\alpha v + \nabla \cdot \rho_\alpha (v_\alpha - v) \tag{1.15}$$

emphasizing the different effects of the flow due to the transport by the bulk motion and of the flux due to molecular diffusion and migration.

If again a separation is made between mean quantities and fluctuations, the non-linear term $\rho_\alpha v_\alpha$ gives two contributions. The first one represents the *advection*, the second one the *dispersion* resulting from the turbulent agitation.

Let $\mathcal{D}_\alpha$ denote the combined effects of turbulent dispersion and molecular diffusion. $\mathcal{D}_\alpha$ can be expressed in terms of the mean concentration by formulas analogous to eqs. 1.12 or 1.13 where new control parameters ("eddy" diffusivities) will appear. In the simplest case:

$$\mathcal{D}_\alpha = \kappa \nabla^2 r_\alpha \tag{1.16}$$

where κ is a control parameter.

Let

$$Q_\alpha = \langle R_\alpha \rangle \tag{1.17}$$

$$I_\alpha = \langle J_\alpha \rangle \tag{1.18}$$

and let $\boldsymbol{\sigma}_\alpha$ denote the mean migration velocity, eq. 1.14 gives:

$$\frac{\partial r_\alpha}{\partial t} + \nabla \cdot r_\alpha u = Q_\alpha + I_\alpha - \nabla \cdot r_\alpha \boldsymbol{\sigma}_\alpha + \mathcal{D}_\alpha \tag{1.19}$$

The roles of the different terms of eq. 1.19 are easily identified:

$\nabla \cdot r_\alpha \boldsymbol{u}$ represents the advection by the mean motion. It introduces a coupling with the mechanical variables.

Q_α represents the local inputs and outputs from the exterior world. (Dumpings are in particular included in Q_α.) Q_α must be given before the model can be operated.

I_α represents the chemical, biochemical and ecological interactions. I_α introduces a coupling between the state variables r_α and interacting state variables r_β, r_γ, etc.

$-\nabla \cdot r_\alpha \boldsymbol{\sigma}_\alpha$ represents the migration. The migration velocities are usually expressed in semi-empirical form inferred from observations, and theoretical reflections on the ecological, chemical and physical properties of the compartments (living species, flocculated suspensions, hydrated combinations, etc.). $\boldsymbol{\sigma}_\alpha$ may be regarded as a control parameter.

$\mathcal{D}_\alpha$ represents the molecular and turbulent dispersion. $\mathcal{D}_\alpha$ is expressed in terms of the mean concentration r_α by semi-empirical

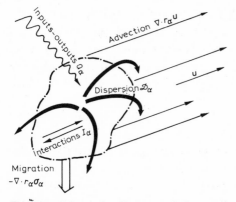

Fig. 1.8. Schematic diagram of the mechanisms affecting the evolution in time of the state variable r_α.

formulas where one or several control parameters ("eddy" diffusivities) are introduced.

The situation is summarized in Fig. 1.8.

(4) The evolution equations for the "thermodynamic" variables are not as easily obtained. For the temperature, at least, it is possible to derive an equation from the energy budget. The procedure, however, involves a series of thermodynamic assumptions which — in addition to being questioned by several specialists of irreversible processes — introduce additional parameters (the specific heat for instance) which, for want of something better, are assumed constant. The equation derived from the energy budget is similar to eq. 1.19 and indeed takes exactly the same form when eq. 1.7 holds (perfectly well-mixed sea, or stratified sea satisfying the Boussinesq approximation). In that case, the mean temperature θ obeys the evolution equation:

$$\frac{\partial \theta}{\partial t} + \nabla \cdot \theta \boldsymbol{u} = Q_\theta + I_\theta + \mathcal{D}_\theta \qquad (1.20)$$

The second term in the left-hand side represents the well-known heat convection, the first two terms in the right-hand side, Q_θ and I_θ, represent the local production of heat respectively by external sources (radiation for instance) and by internal sources (chemical or biochemical reactions for instance). The last term, $\mathcal{D}_\theta$, represents the dispersion of heat by turbulent agitation and molecular diffusion. (The latter is usually called "conduction".)

Similar equations for the diversity or the stability index are still under investigation.

An effort should be made to find or refine the evolution equations for the "thermodynamic" variables which may be extremely helpful in evaluating the state of an ecosystem.

Looking at eqs. 1.4, 1.8 and 1.14, one cannot fail to notice that, although there is an evolution equation for each ρ_α, there are only four for the five unknowns, ρ, p and the three components of the velocity vector. An additional relationship is needed. This is provided by the so-called *equation of state*:

$$\rho = \rho(p,\ \theta,\ \rho_\alpha) \qquad (1.21)$$

This equation is responsible for a coupling between eqs. 1.4 and 1.8 on the one hand, and the system of eq. 1.14 on the other hand.

The equation of state will be discussed in Chapter 2 together with the conditions of applicability of *hydrodynamic models*, restricted to eqs. 1.4 and 1.8 and eventually a limited number of equations of type 1.14.

Finding appropriate evolution equations for the "thermodynamic" variables and an adequate equation of state is one major problem in constructing
a mathematical model. Another one is the determination of the interaction
terms I_α. This involves chemical kinetics and food chain dynamics, and much
work is still needed to identify, formulate analytically and ascertain numerically many interactions. Classical approximations will be presented in Chapter 4 in relation with the discussion of the box or "interaction" models.

1.5. EVOLUTION PRINCIPLES AND CONSTRAINTS

In general the evolution of a system may be subject to evolution principles
(variational principles for instance) and constraints. In some cases, these may
just be another way of expressing the laws of evolution. (The Euler-Lagrange
equations of the variational principles are the same as the evolution equations.) In other cases they may constitute additional requirements which
may bear on control parameters or inputs.

As mentioned before, some of the control parameters are fixed by separate models, experimental data or semi-empirical consideration, some are
accessible to modification by man to conform to some suitably defined
optimal behaviour. These are the guidance parameters to which there correspond guidance principles and constraints.

For the marine system, guidance principles and constraints may arise from
the necessity of preserving valuable resources, limiting pollution, i.e. manag-
ing the system. The management requirements may partly or totally determine the guidance parameters. In addition, they may bring limitations to the
volume or boundary inputs and outputs and find expression in economical
recommendations.

Even when variational principles duplicate the evolution equations, they
may be very helpful. They are often easier to handle numerically and, in the
applications of the model to simulation and prediction, one generally finds a
great advantage in the direct exploitation of variational principles.

For that reason, it is always advisable, when constructing a mathematical
model, to explore the possibility of a variational formulation of the laws of
evolution.

An illustrative example is provided by the Lotka-Volterra model. This is a
completely space-averaged box model describing the evolution with time of
the box-averaged concentrations of interacting "constituents" (chemical substances, biological species, etc.). The fundamental assumption is made that
the laws of interactions can be represented mathematically by quadratic
forms.

Thus, if C_α denotes any state variable (C_α may represent the specific mass of a chemical substance, the specific mass or, rather, the number of individuals of a living species, etc.), Lotka and Volterra postulate an evolution equation of the form:

$$\frac{dC_\alpha}{dt} = a_{\alpha\beta}C_\beta + b_{\alpha\beta\gamma}C_\beta C_\gamma \tag{1.22}$$

where $a_{\alpha\beta}$ and $b_{\alpha\beta\gamma}$ are constants.

Several chemical reactions are described by equations like 1.22. These include: (1) the isomeric transitions $A \to A^*$; (2) the reactions of composition $A + B \to C$; and (3) the reactions of decomposition $A \to B + C$.

Moreover, if one is prepared to introduce the necessary number of intermediate steps involving more or less transitory complexes, one can reduce all chemical reactions to a combination of the three fundamental types above.

The Lotka-Volterra model provides thus a very interesting approach to several aspects of chemical kinetics. The applicability of the model to ecological interactions is clearly demonstrated by considering the example of two species.

The first species is a *prey*. His food supply is assumed unlimited. Hence, his population, measured by the number of individuals n_1, would grow indefinitely without the presence of a *predator* who feeds upon him and without whom he would perish. If n_2 measures the population of the predator, the number of binary encounters effective between the two species is proportional to $n_1 n_2$. Thus:

$$\frac{dn_1}{dt} = e_1 n_1 - d_1 n_1 n_2 \tag{1.23}$$

$$\frac{dn_2}{dt} = -e_2 n_2 + d_2 n_1 n_2 \tag{1.24}$$

where the linear terms represent the growth of n_1 and the decay of n_2 in the absence of interactions, and where, in this particular problem, the coefficients e_1, e_2, d_1, d_2 are positive.

The variational formulation of eq. 1.22 will be shown on the simple system:

$$\frac{dn_1}{dt} = \alpha_1 n_1 + \beta_1 n_1 n_2 \tag{1.25}$$

$$\frac{dn_2}{dt} = \alpha_2 n_1 + \beta_2 n_1 n_2 \tag{1.26}$$

which is equivalent to eq. 1.22 in the case of two state variables only, and include eqs. 1.23 and 1.24 as a special case. The generalization to more than two state variables is conceptually not difficult (e.g. Kerner, 1959).

Eqs. 1.25 and 1.26 admit the particular equilibrium solution:

$$n_1^x = -\frac{\alpha_2}{\beta_2} \tag{1.27}$$

$$n_2^x = -\frac{\alpha_1}{\beta_1} \tag{1.28}$$

if the coefficients α_2 and β_2 on the one hand, and α_1 and β_1 on the other hand are of opposite signs (but not necessarily as in eqs. 1.23 and 1.24).

Assuming that the equilibrium state exists and changing variables to:

$$\mu_1 = \ln\frac{n_1}{n_1^x} \tag{1.29}$$

$$\mu_2 = \ln\frac{n_2}{n_2^x} \tag{1.30}$$

eqs. 1.25 and 1.26 become:

$$\frac{d\mu_1}{dt} = \alpha_1(1-e^{\mu_2}) \tag{1.31}$$

$$\frac{d\mu_2}{dt} = \alpha_2(1-e^{\mu_1}) \tag{1.32}$$

Combining eqs. 1.31 and 1.32, one gets:

$$\frac{d\mu_1}{d\mu_2} = \frac{\alpha_1}{\alpha_2}\frac{1-e^{\mu_2}}{1-e^{\mu_1}} \tag{1.33}$$

and, by integration:

$$\alpha_1(\mu_2-e^{\mu_2})-\alpha_2(\mu_1-e^{\mu_1}) = C^{nt} \tag{1.34}$$

Eq. 1.34 yields thus a first integral of the evolution.

Writing, to conform with classical notations:

$$q = \mu_1 \tag{1.35}$$

$$p = \mu_2 \tag{1.36}$$

$$H = \alpha_1 (p - e^p) - \alpha_2 (q - e^q) \tag{1.37}$$

eqs. 1.31, 1.32 and 1.34 can be put in the *canonical* form:

$$\dot{q} = \frac{\partial H}{\partial p} \tag{1.38}$$

$$\dot{p} = -\frac{\partial H}{\partial q} \tag{1.39}$$

$$\frac{dH}{dt} = 0 \tag{1.40}$$

To canonical equations of this type, one can associate a variational principle:

$$\delta \int_{t_1}^{t_2} L \, dt = 0 \tag{1.41}$$

where L is the Lagrangian.

In general L is a function of q, $\dot{q}$ and t (here t does not appear explicitly because, as a consequence of 1.34, the system is conservative) given by:

$$L = \dot{q} p(q, \dot{q}) - H[q, p(q, \dot{q})] \tag{1.42}$$

Thus, the Lagrangian associated with eqs. 1.35, 1.36 and 1.37 is:

$$L = \alpha_1 \left(1 - \frac{\dot{q}}{\alpha_1}\right) \left[1 - \ln\left(1 - \frac{\dot{q}}{\alpha_1}\right)\right] + \alpha_2 (q - e^q) \tag{1.43}$$

1.6. INITIAL AND BOUNDARY CONDITIONS

The evolution equations must be solved subject to given initial and boundary conditions.

The solution is usually obtained by numerical techniques which imply at one stage or another a discretization in space and time replacing the region under study by a numerical grid of points.

This grid is adapted in each problem to the phenomenon under investigation and the mathematical method used; and the mesh size is chosen from the considerations of the length and time scale of the former and the stability of the latter. There is thus an infinite variety of numerical grids for any mathematical model. These adaptable grids should not be confused with experimental grids of points where samples are taken to determine some initial state of the region or check some results of the model, in most cases, with respect to a particular phenomenon.

These experimental surveys of the region under study are in fact no longer necessary when the mathematical model is sufficiently appropriate and the boundary conditions are provided with sufficient accuracy.

The quality of the model, on the one hand, and the accuracy of the boundary conditions on the other hand, are actually the factors which may limit the success of mathematical modelling (Nihoul, 1973).

Major problems are:

(1) The reliability of data given at open-sea boundaries.

(2) The formulation of boundary conditions at the bottom and at the air—sea interface and the assessment of airborne materials entering the sea.

(3) The identification of coastal inputs (in particular the discharge of pollutants by rivers, sewers, etc.).

The problems arising at open-sea boundaries have been discussed in particular by Nihoul (1973). They will be briefly commented in section 1.8.

The specification of boundary conditions at the air—sea interface and at the sea floor illustrate a typical difficulty of modelling.

As mentioned before, models deal with mean variables (over some characteristic time of interest) and summarize the non-linear effects of the fluctuations in suitably defined "dispersion terms" where several control parameters are introduced. Obviously, the demarcation of the system and, in particular, the definition of its support must conform to this simplification, and the boundaries delimiting the system for a given model are idealized boundaries where most of the intricacy of the real frontiers have been smoothed out by the averaging.

Thus, if $x_3 = \zeta$ and $x_3 = -h$ are the equations of the air—sea interface and the bottom, respectively, and if u denotes the mean velocity vector as defined before, the boundary surfaces will satisfy:

$$\frac{\partial \zeta}{\partial t} + u_1 \frac{\partial \zeta}{\partial x_1} + u_2 \frac{\partial \zeta}{\partial x_2} = u_3 \quad \text{at} \quad x_3 = \zeta \tag{1.44}$$

$$\frac{\partial h}{\partial t} + u_1 \frac{\partial h}{\partial x_1} + u_2 \frac{\partial h}{\partial x_2} = -u_3 \quad \text{at} \quad x_3 = -h \tag{1.45}$$

Eqs. 1.44 and 1.45 express that the "model" boundaries are material surfaces of an idealized fluid moving with the mean velocity. They thus differ from the complicated real frontiers which are jagged material surfaces following the real stirring velocity.

(A particular inference is that problems with different time scales require different model boundaries as, for example, "surface elevation" has a different meaning in tidal models and residual current models.)

Schematically, one might say that each model has its particular conception of the sea surface and of the bottom. Interpreting in mathematical terms the interactions at the sea boundaries is a difficult task. It is complicated in modelling by the obligation to comply with the model's refinement and formulate the boundary conditions with just the right degree of sophistication.

A few examples may serve to illustrate the difficulty*:

(1) Boundary interactions

Specifying boundary conditions appropriate to modelling at the surface and bottom of the sea is a complex problem which "modellers" often explain, with disheartened irony, by the fact that these boundaries just do not exist.

It would be nice indeed to have, as upper and lower frontiers, respectively, a well defined free surface and a non-ambiguous rigid bottom. Unfortunately, none of these boundaries may be taken as entirely free or entirely rigid and, definitely, none can be considered as well defined.

Waves at the sea surface create a first problem. They influence the instantaneous local wind and the subsequent air—sea interactions, but they are themselves (their speed, form and height) function of the wind stress *in the past.* Waves break frequently as a result of interaction between waves of different wave numbers. Wave breaking, which can occur at low wind speed, is more frequent and efficient under strong wind conditions. In the process, spray is produced. Depending on the relative velocities of wind and waves, the disruption of wave crests into spindrift can also produce spray. (At winds above 8 m/sec, there is apparently a sudden increase in the spray load which could be attributed to this mechanism (Krauss, 1967).) Bubbles of air are entrained by the breaking waves. When bubbles burst on reaching the surface, drops are ejected into the air.

An irregular mixing of air and water results on both sides of a jagged interface which one finds difficult to position practically within a complicated zone of transition between the atmosphere and the sea.

---

* The material presented here is borrowed from J.C.J. Nihoul, 1973, Interactions at the sea boundaries as a handicap to modelling. *Proc. Liège Colloq. Ocean Hydrodynamics, 5th, Liège, 1973.*

The situation is similar at the bottom. Sea water contains suspended sediments. Whether the mean turbidity is low (open sea) or high (coastal area), there is inevitably a marked increase in the concentration of sediments as one approaches the sea floor which itself can be permeated with water to some depth. It is almost impossible to draw a line through the solid-water mixture at the bottom of the sea, and the so-called bottom boundary is often nothing more than a conventional limit corresponding to some chosen concentration of the sediments. Moreover, the shear stress exerted by the sea on the bottom may recirculate deposited sediments and erode the sea floor.

Blowing sand (like flying spray) contributes to shade off the border between sea and soil.

A further difficulty in specifying boundary conditions at the sea surface occurs evidently when some alien substance is interposed between air and water. Organic surface-active films are frequently formed on the sea surface. Damping capillary waves, they produce anomalies in the reflection of light which one calls "slicks". Natural sea slicks constituted by monomolecular films of adsorbed organic molecules have been known for a long time. As a result of pollution, contaminant films or layers such as slicks of petroleum products or other chemicals are now more and more often observed.

Surface films modify air—sea interactions, in general in an assuaging way, attenuating small waves, hindering wave formation, reducing air drag, inhibiting gas exchange, etc. (e.g. Garrett, 1972).

Momentum is presumably transferred from wind to waves and, for a substantial part, passed over to the water column by combing waves or other mechanisms (Stewart, 1967; Krauss, 1967). The detailed machinery of this transfer is rather complicated because the wind profile is itself affected by the state of the sea surface, the nature and the shape of the waves, and both wind and waves are largely influenced by the stratification. The complicated interplay between air stability, sea state and wind field reflects on the resulting stress exerted by the wind. Heavy rainfall and wave breaking will increase it. (Momentum transfer is presumably enhanced by the agitation of air caused near the interface by the interpenetration of air and water layers (Tobas and Kunishi, 1970).) Oil slicks decrease the stress. In unstable conditions, a definite augmentation of momentum transfer seems to occur when the wind speed exceeds some 7 m/sec associated with the onset of foaming (De Leonibus, 1971). Acceleration of spray drops (and deceleration of bubbles) produces an additional shear stress which is presumably small at low wind speed but may become relatively more important at higher wind speeds (Krauss, 1967).

The situation is evidently similar for the transport of heat and material across the air—sea interface. Intricated surface mechanisms ensure the transmission of fluxes with variable efficiency. Infrared radiations, for instance, are absorbed at the interface, producing a discontinuity in the heat flux (Krauss, 1967). Pesticides fallout from the atmosphere are often concentrated in sea slicks as a result of their preferential solubility in non-polar organic materials (Seba and Corcoran, 1969). According to MacIntyre (1971) surface-active organic slicks can concentrate all ionic species with high ionic potential relative to Na and Cl. Aerosols produced by the evaporation of spray droplets transfer salt and other substances from the sea to the atmosphere. A substantial amount of volatile hydrocarbons and chemical species which can be concentrated in slicks can be carried into the atmosphere from the sea by this process (Garrett, 1972).

Wave generation can improve significantly the rates of evaporation and aeration. According to Hidy (1972), this must be attributed to the disturbance of the diffusional sublayer near the water boundary, which plays a key role in the material transfer across the naviface.

A similar situation prevails at the bottom. When a viscous layer can be maintained on the sea floor, the flux of material is simply due to the slow sedimentation of particles. When the layer is disrupted by turbulence, the material concentrated in the viscous layer can be recirculated. For still higher values of the turbulent shear, the flow is able to erode the bottom and blow solid sediments into the water column and its turbulent dynamics (McCave, 1970; Cormault, 1971).

A few examples suffice to demonstrate the complexity of the interactions at the upper and lower sea boundaries. One understands the difficulty of specifying judicious boundary conditions appropriate to a model. The uncertainties in the boundary conditions however deeply reflect on the model's predictions, and the inaccuracy of boundary data is often the most severe handicap to modelling.

(2) Boundary conditions

At the air—sea interface, it is generally assumed that the fluxes are proportional to the magnitude of the wind velocity at some reference height (e.g. 10 m). If V, θ, q denote the velocity, the temperature and the moisture (or perhaps the concentration of some contaminant) at the height of reference, and θ_0, q_0 the corresponding surface values, one writes (Krauss, 1972):
(a) momentum flux:

$$\tau_s = \rho V C_{10} \|V\| \tag{1.46}$$

(b) heat flux (divided by the specific heat):

$$h_s = \rho(\theta_0 - \theta) C_{10} \|V\| \tag{1.47}$$

(c) moisture flux:

$$\omega_s = \rho(q_0 - q) C_{10} \|V\|$$ (1.48)

where the "drag coefficient" C_{10} is assumed to be constant.

According to Krauss (1972), "the preceding expressions, applied to good meteorological data, probably can yield flux estimates over the open ocean with a mean expected error of less than 30%". Hidy (1972) anticipates a "more realistic error" as large as 50%. This would seem to be the range in which the data from various observations scatter under comparable wind conditions.

In addition, observations from the Bomex experiment reported by Pond et al. (1971) cast doubt on the validity of eq. 1.47 for the heat transport which is significantly affected by radiation. One may also question the extension of eq. 1.48 to the flux of chemical species which may dissolve or interact preferentially in organic surface films.

At the bottom, an expression similar to 1.46 is generally adopted for the flux of momentum, although, for the convenience of two-dimensional models, one often prefers to express the bottom shear in terms of the depth-averaged velocity u. Thus:

$$\tau_b = D\bar{u}\|\bar{u}\| - m\tau_s$$ (1.49)

where D is a suitable drag coefficient and where the last term in the right-hand side is a small correction introduced to account for the fact that, in case of negligible volume transport of water, there is still a stress exerted by the bottom (Groen and Groves, 1966).

If one assumes, for want of something better, that the heat flux can be derived by analogy with the momentum flux, a different formulation is required for the flux of sediments to account for the recirculation and erosion described above.

If a stable viscous layer exists on the bottom, the turbulent flux vanishes there and the total flux (e.g. b) is reduced to the sedimentation. Thus, one would have, in that case

$$b = -\sigma c_b$$ (1.50)

where σ is the sedimentation velocity and c_b the near-bed concentration.

Eq. 1.50 assumes complete entrapment in the viscous sublayer. To allow for periodic disruption of the layer and ejection of sediments, a correcting factor is introduced, indicating the effective degree of sublayer instability.

One writes (Owen and Odd, 1970):

$$b = -\sigma c_b \left(1 - \frac{\tau_b}{\tau_c}\right)$$ (1.51)

where τ_c is the limiting shear stress above which no deposition takes place.

In eq. 1.51, τ_b is of course variable and the limiting shear stress can be exceeded part of the time (e.g., part of the tidal cycle) still allowing for a positive budget of the sedimentation in the mean.

When the shear stress exceeds a second critical value τ_e, the turbulence is able to erode the sea bed. The flux of eroded material seems to obey the same type of law and Cormault (1971) suggested that it can be expressed by eq. 1.51 where τ_c is replaced by τ_e and σc_b by some constant M depending on the nature of the sea floor.

It is not quite clear if formulas of that type incorporate the action of waves on the sediment transport in the nearshore zone and if expressions like eq. 1.49 for τ_b are accurate enough to be substituted in eq. 1.51 (McCave, 1971).

The boundary conditions discussed briefly in this section have been shown to be often inadequate, always unprecise as a result of the introduction of empirical coefficients C_{10}, D, τ_c, etc., which are control parameters difficult to ascertain experimentally. In fact, too much is being asked from these coefficients. They should integrate the intricated interactions at the sea boundaries with just the right degree of refinement to provide boundary conditions appropriate to the models.

In the present stage, the situation is obviously not very satisfactory and models are often much better than the boundary inputs they elaborate on.

One can however expect a serious improvement in the future years as more and more countries undertake large-scale oceanographic campaigns, participate in international surveys and collaborate to the installation of extended monitoring networks operating from buoys, oceanic platforms, lightvessels, etc. The data acquired continuously by such programs will undoubtedly elucidate considerably the processes occurring at the boundaries and determining the boundary conditions.

In designing such large-scale experiments — and indeed, in general, when calling on the observers for boundary data — it is indispensable to identify and state clearly what information the models require and what accuracy is needed.

As mentioned before, boundaries and boundary conditions must be determined with just the right degree of refinement to conform with the model's sophistication. In addition, the nature and the number of boundary conditions must be adjusted to the necessity of the model. Boundary conditions must be such that the problem set up by the model has a unique solution. A surplus of data can always be used to check the model or the techniques of simulation. A shortage of data hinders the application of the model which can only be exploited after hazardous inter- or extra-polations have supplied the missing information.

One emphasizes here that specifying the required boundary conditions is part of the model's functions. The data to give at the boundaries depend on the scope and the support of the model and also on the characteristic time and length scales of the processes described by the model.

This is not surprising since the evolution equations themselves depend on the scope, support and characteristic scales and the boundary conditions must provide the necessary data to solve the evolution equations in a unique, non-ambiguous way.

The required boundary conditions are thus determined by the final form of the evolution equations, obtained after all simplifications have been made (reduction of support, averaging, etc.) and readily applicable to simulation and prediction.

From the evolution equations there is a classical method to find the boundary conditions appropriate to a given problem. One considers a small domain overlapping the boundary and writes surface sources and interactions as volume contributions with the help of delta functions. One then integrates the evolution equations over the small domain and transforms volume integrals of divergences into surface integrals by Gauss theorem. The contributions of the time derivatives, the fluxes and flows *along* the boundary and the *true volume* sources and interaction terms are proportional to the volume of the integration domain while the contributions from the flows and fluxes *across* the boundary and the *surface* sources and interactions are proportional to the area of the boundary cross-section of the domain. Hence, considering that the domain of integration is small (and indeed goes to zero, from a mathematical point of view), the boundary conditions must express the balance between surface sources and interactions on the one hand and differences in flows and fluxes on the two sides of the boundary on the other hand.

An important remark must be made here. When one reduces the support by integrating over one or several spatial coordinates, one likewise curtails the frontiers of the system. For instance, the marine system sketched in Fig. 1.4 extends over three space dimensions and is limited by *lateral boundaries* (coasts or open-sea boundaries) and by the *air—sea interface* and the *bottom*. If one averages over depth, thus reducing the support to two space dimensions, the simplified system which results has only lateral boundaries. It is on these boundaries that one must give appropriate boundary conditions to solve the, now depth-averaged, evolution equations.

The conditions on the bottom and air—sea surface have not however become obsolete. Indeed they now appear in the evolution equations where they have been introduced by the averaging process.

This will be seen in detail in the following chapter. It can be illustrated simply here by considering the vertical flux of a constituent α. In a three-dimensional model, this flux is represented by a term of the form:

$$\frac{\partial}{\partial x_3}\left(\kappa_3\,\frac{\partial r_\alpha}{\partial x_3}\right) \tag{1.52}$$

where x_3 is the vertical coordinate and κ_3 the vertical coefficient of diffusion.

In a depth-averaged model, this term becomes:

$$H^{-1}\int_{-h}^{\zeta}\frac{\partial}{\partial x_3}\left(\kappa_3\,\frac{\partial r_\alpha}{\partial x_3}\right)\,dx_3 = H^{-1}\left[\left(\kappa_3\,\frac{\partial r_\alpha}{\partial x_3}\right)_\zeta - \left(\kappa_3\,\frac{\partial r_\alpha}{\partial x_3}\right)_{-h}\right] \tag{1.53}$$

where:

$$H = h + \zeta \tag{1.54}$$

is the total depth and where:

$$\left(\kappa_3\,\frac{\partial r_\alpha}{\partial x_3}\right)_\zeta \text{ and } \left(\kappa_3\,\frac{\partial r_\alpha}{\partial x_3}\right)_{-h}$$

denote the surface and bottom values of the flux.

These boundary data have thus been introduced in the evolution equations by the integration over depth.

The same will be true if one averages over more than one spatial variable: boundary inputs and outputs are transformed into "volume" inputs and outputs; they disappear from the boundary conditions but are incorporated in the source-sink terms Q_α (eq. 1.19).

As mentioned before, it is generally difficult to express boundary conditions without having recourse to empirical coefficients. If, by space integration, boundary data are inserted in the evolution equations, they bring along additional control parameters which, in most cases, have a cogent influence on the system's behaviour.

The remark made here enforces the statement that it is the *final* form (all simplifications made) of the evolution equations which sets the *boundary* conditions required to determine the solution, with the understanding that, according to the degree of simplification, exchanges which occur at natural

TABLE 1.2

Résumé of the boundary conditions required for the different types of models

Dimensions of support	Type of model	Conditions required
4		
3 space dimensions + time	time-dependent three-dimensional	initial conditions lateral boundary conditions conditions at air—sea interface and at sea bottom
3		
3 space dimensions	three-dimensional steady state	lateral boundary conditions conditions at air—sea interface and at sea bottom
2 space dimensions + time	time-dependent depth-averaged	initial conditions lateral boundary conditions
2		
2 space dimensions	depth-averaged steady state	lateral boundary conditions
1 space dimension + time	time-dependent depth- and width-averaged	initial conditions up-stream and down-stream boundary conditions
1		
1 space dimension	depth- and width-averaged steady state	up-stream and down-stream boundary conditions
time	completely space-averaged box models	initial conditions

frontiers may not all appear as boundary data but be partly concealed as local inputs or outputs in the equations themselves.

A résumé of the boundary conditions which are required for the different types of models is given in Table 1.2. Initial conditions are indicated when needed.

1.7. THE SIMULATION MODEL

The simulation model is prepared from the mathematical model to be simulated directly on the computer.

The first decision concerns the type of computer which will be used.

In an analog computer the variables are *continuous* d.c. voltages which are *analogous* to the physical variables. All operations are performed *concurrently and in parallel.* The accuracy of an analog simulation is generally not more than two, three or four significant digits. It is limited by the approximation made by each building block in the computer, by uncertainties during computation (noise, non-linear effects, etc.) and by the precision of the output devices (oscilloscope, x—y plotters, etc.).

The variables in a digital computer are *discrete* numbers representing the physical quantities at intervals of the independent variables. The operations are performed *in sequence*. The precision of a digital computer is between six and sixteen digits. The accuracy of a digital simulation is limited by the precision of individual operations (*round-off error*), the programming (the manner in which the operations are sequenced and error is propagated) and the mathematical approximations used for obtaining the discrete-system model (*truncation error*).

A hybrid computer will be essentially constituted by an analog and a digital computer linked together. The combination proves resourceful for complex or fast systems as well as for sophisticated real-time applications. Additional errors arise from sampling analog variables, conversion, reconstruction and round-trip delays of information through the sequential digital computer, and special techniques are needed to compensate for them.

The second step is the programming and the choice of a programming language. This matter is sufficiently well documented to be only mentioned in this chapter.

It is worth noting here that, despite the simplifications which may have been made already in edifying the mathematical model, the construction of the simulation model usually involves many more. Discretization, for the purpose of digital or hybrid computation, for instance, always implies more or less severe approximations. Indeed, the continuous support of the system is then replaced by a discrete grid of points where the state variables are computed by the simulation model. Thus, the micro-system which is enclosed in any one mesh is not described by the simulation model which only considers its whole behaviour as an element of the total system.

In a sense this is similar to what is done in constructing the mathematical model when the state variables are separated into mean parts over some characteristic time of interest and rapidly changing deviations from the mean which are accounted for in some aggregate way only. Indeed, if τ is the characteristic time of separation, one may usually associate to it a characteristic length scale ℓ given by:

$$\ell \sim v\tau \tag{1.54}$$

where v is some velocity characterizing the motion under consideration.

Hence, the separation may be looked at as also cross-ruling the system's support and renouncing a detailed description of fast-varying processes inside meshes of dimensions τ in time and ℓ in space.

There is however an enormous difference between the separation effected in the edification of the mathematical model and the discretization necessitated by the simulation model.

Indeed, while the separation is dictated by the time and length scales of the processes affecting the system, the discretization is guided by independent considerations: (1) the extent of the support and the size of the computers prohibiting minute coverings of the support; and (2) the tractability and the stability of the numerical methods.

A question then commands attention. To what extent are the two procedures, separation in the mathematical model and discretization in the simulation model, compatible?

Although the answer depends upon the particular problem, it is easy to see that a good accord between the two is very unlikely. The reason is that the relationships between the time and length scales characterizing a physical process on the one hand and between the time and space steps chosen, say, to ensure numerical stability, on the other hand, are normally fundamentally different and one cannot expect, in general, similar grids to emerge from the mathematical separation and the numerical discretization.

The mathematical separation has been introduced by considering an averaging time τ sufficiently small for the phenomenon under investigation to pass through the average without significant alteration and sufficiently large to smooth out the details of rapid processes encumbering the analysis.

These rapid processes are often referred to as "fluctuations" because they are essentially produced by erratic motions associated with turbulence and randomly distributed fast oscillations. In most cases, they are even more simply called "turbulent processes" — although, if analyzed in detail, they may not entirely correspond to genuine turbulence (Woods, 1973) — because approximations suggested by the theory of turbulence apply satisfactorily to their description in the mean.

With only an approximate description of this type in mind, one may regard the sea as a hierarchy of turbulent eddies of different length (and time) scales; large eddies permanently breaking down into smaller ones, thus transferring energy to the smallest eddies where viscous dissipation takes place.

Because of the limitations imposed on the vertical components of movement, turbulent eddies in the sea are unlikely to be three-dimensionally isotropic if their scale exceeds a few meters. It is generally admitted however (e.g. Bowden, 1972) that, from a horizontal scale of several hundred kilometers where turbulence is generated by the world-wide wind systems, to a scale of one centimeter where the turbulent energy is dissipated, turbulent eddies are fairly well described by Kolmogorov's theory of locally isotropic turbulence.

According to Kolmogorov, the energy is distributed among different scales of motion according to the law:

$$E(\ell) \sim \epsilon^{2/3} \varrho^{5/3} \tag{1.55}$$

where ℓ is the characteristic length scale of motion, and ϵ is the rate of energy transfer through the descending cascade of eddies.

In the sea however, the energy is not entirely supplied at the outer scale (to the largest eddies). Supply of energy occurs at intermediate scales: from wind currents generated locally, tidal streams, inertial oscillations and waves. It follows that, although eq. 1.55 provides a fair representation of the energy distribution (Bowden, 1972), a different form of 1.55 holds in different bands of scales, each of them corresponding to a different value of ϵ.

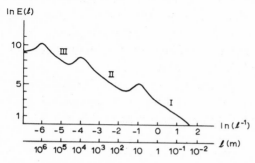

Fig. 1.9. General outline of energy distribution $E(\ell)$ between oceanic motions of different scales (after Ozmidov, 1965).

Fig. 1.9 shows the general outline of energy density distribution as proposed by Ozmidov (1965). Comparatively, few sets of observations are available from which the values of the parameter ϵ could be obtained. Tentative orders of magnitude estimated from the current literature are given in Table 1.3 merely to have figures available for the present discussion. One emphasizes that the modification of the reference values of ϵ may slightly change the orders of magnitude computed below but not the spirit of the conclusions drawn from them.

In accordance with eq. 1.55, the characteristic velocity and time scale associated with "eddies" of size ℓ can be estimated as:

$$v_\varrho \sim \epsilon^{1/3} \varrho^{1/3} \tag{1.56}$$

$$\tau_\varrho \sim \epsilon^{-1/3} \varrho^{2/3} \tag{1.57}$$

Hence, the average over a given characteristic time τ smoothes out the sea motions whose length scale is less than:

$$\ell \sim \epsilon^{1/2} \tau^{3/2} \tag{1.58}$$

TABLE 1.3

Tentative estimates of the rate of energy transfer in
different ranges of the length scale ℓ

Length scales ℓ (m)	Rate of energy transfer ϵ (m^2/sec^3)
range I	10^{-6}
range II	10^{-8}
range III	10^{-10}

The water agitation produced by the small-scale motions cannot accordingly
be described in detail. Only its cumulative effects through the non-linear
terms are incorporated in the model and interpreted as an enhanced dispersion of momentum, heat, contaminants, etc.

As mentioned before, the turbulent dispersion is represented in the evolution equations by terms similar to the molecular diffusion terms but with
much larger diffusion coefficients. According to Kolmogorov's theory, these
"eddy" diffusion coefficients can be estimated in terms of the characteristic
time τ and length ℓ of separation by formulas of the type:

$$\nu \sim \epsilon^{1/3} \ell^{4/3} \sim \epsilon \tau^2 \tag{1.59}$$

For instance, a tidal model of the North Sea is concerned with a phenomenon whose characteristic time scale T is the main tidal period (e.g., a time of
the order of 12 h). Rapidly varying water movements (e.g., of characteristic
time less than a minute) are obviously not directly relevant to the model
which may be content with some aggregate estimation of their dispersive
action.

Hence, it would seem that choosing an intermediate time of, e.g., 5 min
the averaging would not affect the tidal oscillations significantly but would
eliminate the cumbrous details of the velocity field.

A typical length scale of the order of 5 m can be associated by eq. 1.58 to
the characteristic time of 5 min.

The action of turbulent eddies of sizes smaller than 5 m will be expressed
in the evolution equations by a dispersion term controlled by a dispersion
coefficient ν which, estimated by Kolmogorov's theory in first approximation, would be of the order of 10^{-1} m^2/sec.

Now it would be preposterous in the present stage of computing facilities
to attempt a simulation of the tidal oscillations of the North Sea with a
spatial grid of a 5-m mesh size. The spacing between grid points covering the
entire North Sea will typically be of the order of kilometers (e.g., 5 km).

As the simulation model cannot go into the details of motions whose

TABLE 1.4

Characteristic length and time scales and "eddy diffusivities" for a tidal model of the North Sea (didactical example)

Mathematical model	Simulation model
I. Separation between "mean" motion and "turbulence" effected at $\ell \sim 5$ m $\tau \sim 300$ sec	I. Dimensions of grid mesh in space and time $\ell \sim 5 \cdot 10^3$ m $\tau_1 \sim 60$ sec
II. Eddy viscosity $\nu \sim 10^{-1}$ m^2/sec	II. Numerical viscosity $\nu \sim 10^2$ m^2/sec
	III. Characteristic time of motions of scale ℓ $\tau_2 \sim 10^5$ sec

characteristic scales are smaller than the mesh size, the discretization effectively introduces a separation at the scale of 5 km. The eddy diffusivity necessary to account for all smaller-scale motions is then of the order of 10^2 m^2/sec.

The characteristic time associated with scales of motions of the order of a few kilometers is, estimated from eq. 1.57, of the order of a day.

In addition to being physically meaningless since the tidal period is about half a day, a time interval of a day cannot be used to discretize the time axis. Numerical stability requires indeed much smaller time steps. For a spatial mesh size of 5 km, for instance, typical time steps would be of the order of 1 min (e.g. Fischer, 1959, 1965; Kasahara, 1965; Harris and Jelesnianski, 1965; Ronday, 1972).

A table of the different orders of magnitude (Table 1.4) reveals the differences between the separation in the mathematical model between "mean" motion and "turbulence" and the discretization in the simulation model.

Although this is inevitable in most cases, one can see that the simulation model sacrifices much of the physical insight of the mathematical model. The high time resolution of the simulation model is indeed delusory as small time-scale processes are smoothed out by the coarse spatial resolution. In addition, interpreting all motions of smaller scales than the mesh size as a dispersing agitation becomes hazardous when the size of the mesh increases too much. One can object that turbulence theory and eddy diffusivity concepts cannot be extended that far. The numerical viscosity is then rather a computing artifice with few roots in nature.

The several orders of magnitude differences between length and diffusivity scales in the mathematical model and the simulation model is a source of many simulation hazards and misinterpretation.

REFERENCES

Adam, Y.A., 1973. Nato Science Committee Conference on Modelling of Marine Systems, Ofir, Portugal, 1973.

Bowden, K.F., 1972. In: *Proc. Liège Colloq. Ocean Hydrodynamics, 3rd, Liège, 1971. Mém. Soc. R. Sci. Liège*, 2: 67.

Cormault, P., 1971. *Proc. Congr. I.A.H.R., 14th, Paris, 1971.*

De Leonibus, P.S., 1971. *J. Geophys. Res.*, 76: 6506.

Fischer, G., 1959. *Tellus*, 11: 60.

Fischer, G., 1965. *Mon. Weather Rev.*, 93: 1.

Garrett, W.D., 1972. In: D. Dyrssen and D. Jagner (Editors), *The Changing Chemistry of the Ocean*. Wiley, New York, N.Y., pp. 75—91.

Groen, P. and Groves, G.W., 1966. Surges. In: M.N. Hill (Editor), *The Sea*. Interscience, New York, N.Y., pp. 611—646.

Harris, D.L. and Jelesnianski, C.P., 1965. *Mon. Weather Rev.*, 92: 409.

Hidy, G.M., 1972. *Bull. Am. Meteorol. Soc.*, 53: 1083.

Kasahara, A., 1965. *Mon. Weather Rev.*, 93: 27.

Kerner, E.H., 1959. *Bull. Math. Biophys.*, 21: 93.

Krauss, E.B., 1967. *Adv. Geophys.*, 12: 213.

Krauss, E.B., 1972. *Atmosphere—Ocean Interaction*. Clarendon Press, Oxford, 275 pp.

MacIntyre, F., 1971. *Tellus*, 22: 451.

McCave, I.N., 1970. *J. Geophys. Res.*, 75: 4151.

McCave, I.N., 1971. *J. Sediment. Petrol.*, 41: 89.

Nihoul, J.C.J., 1972. In: *Proc. Liège Colloq. Ocean Hydrodynamics, 3rd, Liège, 1971. Mém. Soc. R. Sci. Liège*, 2: 111.

Nihoul, J.C.J., 1973. In: E. Goldberg (Editor), *North Sea Science Book*. M.I.T. University Press, Cambridge, Mass., pp. 43—57.

Owen, M.W. and Odd, N.V.M., 1970. *Proc. Int. Conf. Utilization of Tidal Power, Halifax, N.S.* Department of Energy, Mines and Resources, Ottawa, Ont.

Ozmidov, R.V., 1965. *Bull. Acad. Sci. U.S.S.R., Atmos. Oceanic Phys. Sci.*, 1: 439.

Pichot, G., 1973. Nato Science Committee Conference on Modelling of Marine Systems, Ofir, Portugal, 1973.

Pond, S., Phelps, G., Paquin, J.E., McBean, G. and Stewart, R.W., 1971. *J. Atmos. Res.*, 28: 901.

Ronday, F.C., 1972. Modèle mathématique pour l'étude de la circulation résiduelle dans la mer du Nord. *Marine Sciences Branch, Manuscript Report Series*, 27 (Ottawa).

Seba, D.B. and Corcoran, E.F., 1969. *Pestic. Monit. J.*, 3: 190.

Steele, J.M., 1958. *Rapp. P.-V. Réun. Cons. Perm. Int. Explor. Mer.*, 144.

Stewart, R.W., 1967. *Phys. Fluids Suppl.*, 10: 547.

Tobas, Y. and Kunishi, H., 1970. *J. Ocean Soc. Jap.*, 26: 71.

Woods, J.D., 1973. *Proc. Liège Colloq. Ocean Hydrodynamics, 5th, Liège, 1973.*

CHAPTER 2

HYDRODYNAMIC MODELS

Jacques C.J. Nihoul

2.1. INTRODUCTION

The size of a mathematical model depends on its scope. A comprehensive description of a marine system obviously requires a great number of state variables and evolution equations, and considerable models must be constructed to provide a predictive knowledge of all important aspects of the system. The treatment of a large model however is generally fairly complicated. Large computers and often particularly skilful numerical techniques are necessary. The computing time can be prohibitively large. Moreover, as the size of the model increases, its formulation requests more empirical data and experimentally guided assumptions to describe the spreading web of interactions and fix appropriate values of the multiple parameters. The reliability of the model's predictions may be affected.

For that reason, as mentioned in the introduction of this book, in the same time as one works for the development and perfection of better and better multi-purpose models, one derives, with well-tried elements extracted from the successively improved versions of these models, simpler models, more limited in scope, but amenable to speedy and reliable exploitation.

Hydrodynamic models are models of this type. Their objective is the study of the mechanical variables (cf. Fig. 1.5). They seek a preliminary description of the velocity field (tides, wind currents, etc.), an understanding of the sea hydrodynamics.

The subject is obviously extremely important by itself. Prediction of storm surges and tides, surface elevation, current patterns, are all essential factors of navigation, fishing, protection of the coasts, etc. In addition, the whole dynamics of the marine system is driven by the sea motion. Dispersion of nutrients or pollutants, migrations, sedimentation and bottom erosion are conditioned by the mixing and circulation of the water masses.

The determination of the mechanical variables would thus seem to be the first step in any effort to model a marine system. Knowing the velocity field, it can be substituted in equations like 1.19 or 1.20 to determine the distribution of different constituents or temperature.

The feasibility of a purely hydrodynamic model, however, depends strongly on the equation of state (eq. 1.21) and the possibility of decoupling between eqs. 1.4 and 1.10 on the one hand and the system of eqs. 1.19 and 1.20 on the other hand.

The most evident case is that of a perfectly well-mixed sea where one can assume:

$$\rho = C^{nt} \tag{2.1}$$

Eq. 2.1 is the simplest form of eq. 1.21.

The velocity field is then completely described by eqs. 1.7 and 1.10. The solution of these equations can be later substituted in 1.19 and 1.20 which can, in turn, be solved for r_α and θ provided the right-hand side is known and, in particular, when the constituent is passive ($I_\alpha = 0$) and its evolution is not coupled to other constituents (r_β). Dispersion models of this type will be examined in the next chapter.

In the hierarchy of difficulties, one can consider next hydrodynamic models where ρ is: (a) a function of salinity or temperature; (b) a function of salinity and temperature; (c) a function of salinity, temperature and turbidity.

In general, eqs. 1.6 and 1.7 hold. Eq. 1.10 is then coupled with a limited number of equations of type 1.19 or 1.20 (equations for the salinity, temperature, turbidity, etc.) through the specific mass ρ in the body force terms.

An intermediate approximation consists in considering the sea as a superposition of two or more homogeneous layers of constant but different densities. Eq. 2.1 can then be applied in each layer which can be modelled separately taking into account the additional inputs and outputs representing exchanges between adjacent layers.

2.2. DEPTH-AVERAGED MODELS

Almost all existing hydrodynamic models are depth-averaged three-dimensional models (two space dimensions and time).

Four-dimensional models have been seldom attempted yet although recently Heaps (1972) suggested a method — basically replacing straight integration by integral transform over depth — by which in principle the depth variations could be recovered at the end.

(1) Depth-averaged models rely heavily on the so-called quasi-static approximation by which the pressure can easily be eliminated from the averaged equations. In most cases, the density of sea water ρ is assumed constant.

Then, in rectangular coordinates x_1, x_2, x_3 where the x_3-axis is vertical upwards, the vertical component of eq. 1.10 reads:

$$\frac{\partial u_3}{\partial t} + 2(\Omega_1 u_2 - \Omega_2 u_1) + \nabla \cdot uu_3 = -\frac{1}{\rho}\frac{\partial p}{\partial x_3} - g + \mathcal{D}_3 \tag{2.2}$$

(It is assumed that the only important contribution to the vertical component of the volume force $\langle F \rangle$ is gravity; Ω_1 and Ω_2 are the horizontal components of $\mathbf{\Omega}$.)

The terms in the left-hand side of eq. 2.2 represent the acceleration of the vertical motion. This acceleration is generally negligible as compared to the acceleration of gravity which is of the order of 10 m/sec^2. If the horizontal velocity is of the order of 1 m/sec, the Coriolis term is 10^5 smaller than the gravity term. Observations in different sea areas show that the dispersion term $\mathcal{D}_3$ is also much smaller than g.

If, for instance, the estimates of section 1.7 (Chapter 1) are used, one finds, for a velocity of order 1 m/sec:

$$\frac{\mathcal{D}_3}{g} \sim \frac{\nu u}{\ell^2 g} \sim 4 \cdot 10^{-4}$$

Hence, the gravity force can only be balanced by the pressure gradient and, in good approximation:

$$\frac{\partial p}{\partial x_3} = -\rho g \tag{2.3}$$

Integrating over x_3, one obtains:

$$p = -\rho g x_3 + f(x_1, x_2) \tag{2.4}$$

At the free surface, $x_3 = \zeta$, the pressure must be equal to the atmospheric pressure p_a. Hence:

$$p_a = -\rho g \zeta + f(x_1, x_2) \tag{2.5}$$

This determines the so-far arbitrary function f. Combining eq. 2.4 and 2.5, one gets:

$$p = p_a + \rho g \zeta - \rho g x_3 \tag{2.6}$$

(2) The horizontal components of the Coriolis acceleration are:

$$2\Omega_2 u_3 - f u_2 \tag{2.7}$$

$$-2\Omega_1 u_3 + f u_1 \tag{2.8}$$

where the vertical component of the rotation vector $\boldsymbol{\Omega}$ has been written:

$$\Omega_3 \equiv \tfrac{1}{2} f \tag{2.9}$$

to conform to usual notations.

It is reasonable to assume that the vertical velocity u_3 is much smaller than the horizontal velocity. Hence the first terms may be neglected in eqs. 2.7 and 2.8.

Combining eqs. 1.10, 2.1, 2.6, 2.7 and 2.8 and approximating the dispersion term by eq. 1.12, one has then, for the horizontal components of the velocity vector:

$$\frac{\partial u_1}{\partial t} + \frac{\partial}{\partial x_1}(u_1^2) + \frac{\partial}{\partial x_2}(u_1 u_2) + \frac{\partial}{\partial x_3}(u_1 u_3) - fu_2$$

$$= -\frac{\partial}{\partial x_1}\left(\frac{p_a}{\rho} + g\zeta\right) + \xi_1 + \nu\left(\frac{\partial^2 u_1}{\partial x_1^2} + \frac{\partial^2 u_1}{\partial x_2^2} + \frac{\partial^2 u_1}{\partial x_3^2}\right) \tag{2.10}$$

$$\frac{\partial u_2}{\partial t} + \frac{\partial}{\partial x_1}(u_1 u_2) + \frac{\partial}{\partial x_2}(u_2^2) + \frac{\partial}{\partial x_3}(u_2 u_3) + fu_1$$

$$= -\frac{\partial}{\partial x_2}\left(\frac{p_a}{\rho} + g\zeta\right) + \xi_2 + \nu\left(\frac{\partial^2 u_2}{\partial x_1^2} + \frac{\partial^2 u_2}{\partial x_2^2} + \frac{\partial^2 u_2}{\partial x_3^2}\right) \tag{2.11}$$

where ξ_1 and ξ_2 are the horizontal components of the volume force $\langle F \rangle$ and ν the eddy viscosity assumed constant.

(3) The depth-averaged motion is described in terms of the mean velocity $\bar{u}$ or the total flow-rate U defined by:

$$U = U_1 e_1 + U_2 e_2 = H\bar{u} \tag{2.12}$$

$$U_i = \int_{-h}^{\zeta} u_i \, dx_3 \tag{2.13}$$

where H is the total depth, i.e.:

$$H = h + \zeta \tag{2.14}$$

Deviations from the vertical mean are indicated by $\hat{}$. Thus:

$$u_i = \bar{u}_i + \hat{u}_i \tag{2.15}$$

with

$$\int_{-h}^{\zeta} \hat{u}_i \, dx_3 = 0 \qquad (2.16)$$

Integrals over x_3 of partial derivatives with respect to t, x_1 or x_2 can be expressed in terms of the derivatives of the integrals by the formula:

$$\int_{-h}^{\zeta} \frac{\partial f}{\partial \eta} \, dx_3 = \frac{\partial}{\partial \eta} \int_{-h}^{\zeta} f \, dx_3 - f(\zeta) \frac{\partial \zeta}{\partial \eta} - f(-h) \frac{\partial h}{\partial \eta} \qquad (2.17)$$

where η stands for any of the variables t, x_1 and x_2 and where f is any function of t, x_1, x_2 and x_3, $f(\zeta) \equiv f(t, x_1, x_2, \zeta)$ and $f(-h) \equiv f(t, x_1, x_2, -h)$ its values at the sea surface and at the sea floor, respectively.

(4) In rectangular coordinates, eq. 1.7 reads:

$$\frac{\partial u_1}{\partial x_1} + \frac{\partial u_2}{\partial x_2} + \frac{\partial u_3}{\partial x_3} = 0 \qquad (2.18)$$

Integrating over depth, one obtains:

$$\int_{-h}^{\zeta} \left(\frac{\partial u_1}{\partial x_1} + \frac{\partial u_2}{\partial x_2} \right) dx_3 + u_3(\zeta) - u_3(-h) = 0 \qquad (2.19)$$

Commuting the integration and derivation signs according to eq. 2.17 and eliminating $u_3(\zeta)$ and $u_3(-h)$ by eqs. 1.44 and 1.45, once can write eq. 2.19 in the form:

$$\frac{\partial H}{\partial t} + \frac{\partial U_1}{\partial x_1} + \frac{\partial U_2}{\partial x_2} = 0 \qquad (2.20)$$

where H is given by eq. 2.14 and:

$$\frac{\partial H}{\partial t} \sim \frac{\partial \zeta}{\partial t} \qquad (2.21)$$

(neglecting slow time deformations of the bottom).

Equivalent forms of eq. 2.20 are:

$$\frac{\partial H}{\partial t} + \frac{\partial}{\partial x_1} (H\bar{u}_1) + \frac{\partial}{\partial x_2} (H\bar{u}_2) = 0 \qquad (2.22)$$

$$\frac{\partial H}{\partial t} + \bar{u} \cdot \nabla H + H \nabla \cdot \bar{u} = 0 \qquad (2.23)$$

where ∇ reduces to:

$$\nabla = e_1 \frac{\partial}{\partial x_1} + e_2 \frac{\partial}{\partial x_2} \tag{2.24}$$

since the functions H, U and $\bar{u}$ do not depend on x_3.

One notes that, even if the divergence of the three-dimensional velocity field $\bar{u}$ is assumed zero, the divergence of the two-dimensional mean field $\bar{u}$ is not zero.

However if ζ is everywhere negligible with respect to h, and if h varies slowly in space as compared to $\bar{u}$ and ζ, eq. 2.23 can be approximated by:

$$\nabla \cdot \bar{u} = 0 \tag{2.25}$$

Indeed if L is the characteristic length of variation of h and ℓ, the characteristic length of variation of ζ and $\bar{u}$, the first two terms of eq. 2.23, can be estimated as:

$$\frac{\partial H}{\partial t} \sim \frac{\partial \zeta}{\partial t} \sim O\left(\frac{\zeta \bar{u}}{\ell}\right)$$

$$\bar{u} \cdot \nabla H \sim \bar{u} \cdot \nabla \zeta + \bar{u} \cdot \nabla h \sim O\left(\frac{\zeta \bar{u}}{\ell}\right) + O\left(\frac{\bar{u} h}{L}\right)$$

while the third term is the sum of two parts, each of the order:

$$H \frac{\partial \bar{u}_i}{\partial x_i} \sim O\left(\frac{h \bar{u}}{\ell}\right)$$

If one can assume

$$\ell \ll L ; \quad \zeta \ll h$$

the contributions from the first two terms are all negligible as compared to the contributions from the third term. These must consequently balance each other and eq. 2.25 holds.

(5) The integration of eq. 2.1 presented no difficulty. The equation being linear, all terms containing deviations $\hat{u}_i$ from the mean field disappeared in the integration process by virtue of eq. 2.16. The same is not true for the momentum equation which contains products of the velocity vector components. Indeed, one finds:

$$H^{-1} \int_{-h}^{\zeta} u_i u_j \, dx_3 = \bar{u}_i \bar{u}_j + H^{-1} \int_{-h}^{\zeta} \hat{u}_i \hat{u}_j \, dx_3 \tag{2.26}$$

Thus, the average of a product gives two contributions. The first one is the product of the means, the second one the mean product of the deviations.

This is similar to the situation discussed in section 1.4 (Chapter 1) when the actual velocity vector was separated into a mean part (over a characteristic time τ) and a fluctuation around the mean. Although the average then was a time average while in eq. 2.26 it is a depth average, the same difficulty occurs: when averaging the evolution equations, all terms cannot be expressed in terms of average quantities; the non-linear terms also contain average products of fluctuations or deviations.

These non-linear contributions must somehow be expressed in terms of the mean variables if one wishes to proceed with the model without undertaking a detailed study of the fluctuations or deviations.

Acknowledging the essentially dispersive action of the fluctuations, their general effect on the mean motion has been shown in Chapter 1 to be equivalent to a diffusion term similar to the molecular diffusion term but with much larger diffusion coefficients.

One may argue that the mean product of deviations occurring in eq. 2.16 has a similar effect; differences in local velocities contributing to the spreading of momentum and, as discussed in the next chapter, to the dispersion of temperature, nutrients, pollutants, etc.

This effect has been called the *shear effect* because deviations would be zero if the velocity were uniform over the depth, and terms like the second term in the right-hand side of eq. 2.26 only exist because there is a vertical gradient of velocity or "shear".

The shear effect will be discussed in detail in the next chapter. It will be shown that it plays an essential part in the dispersion of contaminants, and it will be necessary to find appropriate formulas to express it in terms of the depth mean variables.

In hydrodynamic models, one usually makes crude approximations which amount more or less to including the shear effect in the dispersion term and adjusting the coefficient of dispersion accordingly.

Indeed, one can argue that hydrodynamic models of large sea regions can be less delicate in this respect than dispersion models concerned with the spreading of a contaminant over an area of a few kilometers around the point of release. The coarseness of the grid imposed to the simulation model by the extent of the modelled region and the subsequent approximations associated with the numerical viscosity coefficients suggest that the formulation is sufficiently rudimental, hence flexible, to incorporate the shear effect.

It is important to note, however, that including the shear effect may notably change the magnitude of the dispersion coefficients.

Indeed, if a shear effect diffusivity ν_s is introduced such that:

$$H^{-1} \int_{-h}^{\zeta} \hat{u}_i \hat{u}_j \, dx_3 = \nu_s \frac{\partial \bar{u}_i}{\partial x_j} \tag{2.27}$$

comparing the orders of magnitude of the two sides of the equation, one finds that ν_s must be of the order:

$$\nu_s \sim \frac{\ell \hat{u}^2}{\bar{u}} \tag{2.28}$$

where $\bar{u}$ and $\hat{u}$ are typical values of the mean and deviation velocities and ℓ the characteristic length of horizontal variations.

The ratio $\hat{u}^2/\bar{u}$ depends on the vertical distribution of u. It would be very small if the velocity was nearly uniform over the depth. One can argue however that this cannot be the case because the velocity must be maximum at the surface and zero at the bottom, and $\hat{u}$ which is necessarily $\sim \bar{u}$ at the bottom could be reasonably important over a large part of the water column. Thus $\hat{u}^2/\bar{u}$ could be an appreciable fraction of $\bar{u}$.

If this is the case, the coefficient ν_s can be one or two orders of magnitude larger than the eddy viscosity ν_t which can be estimated by:

$$\nu_t \sim \ell v_\varrho \tag{2.29}$$

where the characteristic velocity:

$$v_\varrho \sim \epsilon^{1/3} \varrho^{1/3}$$

associated with the eddies of size ℓ is considerably smaller than $\bar{u}$.

The combined effect of shear and turbulence is accounted for in most models by introducing a new dispersion coefficient a such that:

$$\int_{-h}^{\varsigma} \nu \left(\frac{\partial^2 u_i}{\partial x_1^2} + \frac{\partial^2 u_i}{\partial x_2^2} \right) dx_3 - \sum_{j=1}^{2} \frac{\partial}{\partial x_j} \int_{-h}^{\varsigma} \hat{u}_i \hat{u}_j \, dx_3 = a \left(\frac{\partial^2 U_i}{\partial x_1^2} + \frac{\partial^2 U_i}{\partial x_2^2} \right) \tag{2.30}$$

when one works with the flow rate U, or:

$$H^{-1} \int_{-h}^{\varsigma} \nu \left(\frac{\partial^2 u_i}{\partial x_1^2} + \frac{\partial^2 u_i}{\partial x_2^2} \right) dx_3 - H^{-1} \sum_{j=1}^{2} \frac{\partial}{\partial x_j} \int_{-h}^{\varsigma} \hat{u}_i \hat{u}_j \, dx_3$$

$$= a \left(\frac{\partial^2 \bar{u}_i}{\partial x_1^2} + \frac{\partial^2 \bar{u}_i}{\partial x_2^2} \right) \tag{2.31}$$

when one works with the mean velocity $\bar{u}$.

The coefficient a has obviously not exactly the same meaning in eqs. 2.30 and 2.31, as the passage from one to the other would involve space derivatives of the total depth H. However, in view of the approximate character of such formulations, one generally agrees that refinements would be illusory and that, according to the terminology of the problem, one will use formula

2.30 or 2.31, considering a as a control parameter to be determined empirically.

(6) The left-hand sides of eqs. 2.30 and 2.31 do not contain contributions from the second derivatives of the velocity components with respect to x_3. These contributions are singled out in the integrated equations. Indeed:

$$\nu \frac{\partial^2 u_1}{\partial x_3^2} = \frac{\partial}{\partial x_3} \left(\nu \frac{\partial u_1}{\partial x_3} \right)$$

and

$$\nu \frac{\partial^2 u_2}{\partial x_3^2} = \frac{\partial}{\partial x_3} \left(\nu \frac{\partial u_2}{\partial x_3} \right)$$

are exact differentials. Integrating over depth, one obtains:

$$\int_{-h}^{\zeta} \nu \frac{\partial u_i}{\partial x_3^2} \, dx_3 = \left(\nu \frac{\partial u_i}{\partial x_3} \right)_{z=\zeta} - \left(\nu \frac{\partial u_i}{\partial x_3} \right)_{z=-h} \qquad i = 1,2 \qquad (2.32)$$

One particularizes here the general remark made in Chapter 1 that integration over space incorporates some of the boundary conditions in the evolution equations. The boundary terms in the right-hand side of eq. 2.32, if multiplied by the specific mass ρ, represent the components of the so-called surface and bottom stresses, noted $\boldsymbol{\tau}_s$ and $\boldsymbol{\tau}_b$ in section 1.6 and associated with the flux of momentum at the air—sea interface and at the bottom.

Hence, using eqs. 1.46 and 1.49 and setting:

$$C = C_{10}(1 + m)$$

for brevity, one can write:

$$\int_{-h}^{\zeta} \nu \frac{\partial^2 u_i}{\partial x_3^2} \, dx_3 = \rho^{-1}(\tau_{s,i} - \tau_{b,i})$$

$$\qquad (2.33)$$

$$= CV_i \|V\| - \frac{D}{H^2} U_i \|U\| \qquad i = 1,2$$

where V is the wind velocity at the reference height.

(7) Integrating the two components of the momentum eq. 1.10 and using 1.44, 1.45, 2.17, 2.30 and 2.33, one obtains:

$$\frac{\partial U_1}{\partial t} + \frac{\partial}{\partial x_1}\left(\frac{U_1^2}{H}\right) + \frac{\partial}{\partial x_2}\left(\frac{U_1 U_2}{H}\right) - fU_2$$

$$= H\left[\bar{\xi}_1 - \frac{\partial}{\partial x_1}\left(\frac{p_a}{\rho} + g\zeta\right)\right] + a\left(\frac{\partial^2 U_1}{\partial x_1^2} + \frac{\partial^2 U_1}{\partial x_2^2}\right) + CV_1\|V\| - \frac{D}{H^2}U_1\|U\|$$

$$(2.34)$$

$$\frac{\partial U_2}{\partial t} + \frac{\partial}{\partial x_1}\left(\frac{U_1 U_2}{H}\right) + \frac{\partial}{\partial x_2}\left(\frac{U_2^2}{H}\right) + fU_1$$

$$= H\left[\bar{\xi}_2 - \frac{\partial}{\partial x_2}\left(\frac{p_a}{\rho} + g\zeta\right)\right] + a\left(\frac{\partial^2 U_2}{\partial x_1^2} + \frac{\partial^2 U_2}{\partial x_2^2}\right) + CV_2\|V\| - \frac{D}{H^2}U_2\|U\|$$

$$(2.35)$$

or using eqs. 2.12, 2.20 and 2.31 instead of 2.30:

$$\frac{\partial \bar{u}_1}{\partial t} + \bar{u}_1 \frac{\partial \bar{u}_1}{\partial x_1} + \bar{u}_2 \frac{\partial \bar{u}_1}{\partial x_2} - f\bar{u}_2$$

$$(2.36)$$

$$= \bar{\xi}_1 - \frac{\partial}{\partial x_1}\left(\frac{p_a}{\rho} + g\zeta\right) - \frac{D}{H}\bar{u}_1\|\bar{u}\| + a\left(\frac{\partial^2 \bar{u}_1}{\partial x_1^2} + \frac{\partial^2 \bar{u}_1}{\partial x_2^2}\right) + \frac{C}{H}V_1\|V\|$$

$$\frac{\partial \bar{u}_2}{\partial t} + \bar{u}_1 \frac{\partial \bar{u}_2}{\partial x_1} + \bar{u}_2 \frac{\partial \bar{u}_2}{\partial x_2} + f\bar{u}_1$$

$$(2.37)$$

$$= \bar{\xi}_2 - \frac{\partial}{\partial x_2}\left(\frac{p_a}{\rho} + g\zeta\right) - \frac{D}{H}\bar{u}_2\|\bar{u}\| + a\left(\frac{\partial^2 \bar{u}_2}{\partial x_1^2} + \frac{\partial^2 \bar{u}_2}{\partial x_2^2}\right) + \frac{C}{H}V_2\|V\|$$

Eqs. 2.20, 2.34 and 2.35 constitute the fundamental system of evolution equations for the three unknowns H, U_1 and U_2. Alternatively, eqs. 2.23, 2.36 and 2.37 can be solved for H and the mean field $\bar{u}$.

The basic concepts and fundamental equations of a depth-averaged hydrodynamic model are summarized in vectorial form in Table 2.1.

It is instructive to recall the specific physical significance of each term in the equations of Table 2.1.

Considering the equations for the mean field $\bar{u}$ (the description is easily transposable to the flow rate U), one sees that the time variation of the velocity vector is the result of

α advection $\bar{u} \cdot \nabla \bar{u}$

β rotation $fe_3 \wedge \bar{u}$ produced by the Coriolis effect in axes fixed on the rotating earth

TABLE 2.1

Basic concepts and equations of a depth-averaged hydrodynamic model

---

$$U = \int_{-h}^{\zeta} U \, dx_3 = H \, \bar{u}$$

$$H = h + \zeta$$

A. $\dfrac{\partial H}{\partial t} + \nabla \cdot U = 0$ 　　　　(continuity equation)

$$\frac{\partial U}{\partial t} + \nabla \cdot (H^{-1} \, UU) + f e_3 \wedge U = H \left[\bar{\xi} - \nabla \left(\frac{p_a}{\rho} + g\zeta \right) \right] + a \, \nabla^2 U - \frac{D}{H^2} U \| U \| + C \, V \| V \|$$

B. $\dfrac{\partial H}{\partial t} + \bar{u} \cdot \nabla H + H \, \nabla \cdot \bar{u} = 0$

$$\frac{\partial \bar{u}}{\partial t} + \bar{u} \cdot \nabla \bar{u} + f e_3 \wedge \bar{u} = \bar{\xi} - \nabla \left(\frac{p_a}{\rho} + g\zeta \right) + a \, \nabla^2 \bar{u} - \frac{D}{H} \bar{u} \| \bar{u} \| + \frac{C}{H} V \| V \|$$

$$\bar{\xi} = \bar{\xi}_1 \, e_1 + \bar{\xi}_2 \, e_2$$

$$\nabla = e_1 \frac{\partial}{\partial x_1} + e_2 \frac{\partial}{\partial x_2}$$

γ　　mixing by shear effect and turbulence $a\nabla^2 u$

δ　　friction on the bottom $-(D/H)\bar{u} \| \bar{u} \|$

ϵ　　acceleration by external agents of three different types

$\epsilon.1$　the wind stress on the sea surface $(C/H)V \| V \|$ related to the wind velocity V at some reference height

$\epsilon.2$　the gradient of the atmospheric pressure and of the surface elevation $-\nabla \left[(p_a/\rho) + g\zeta \right]$

$\epsilon.3$　the external force $\bar{\xi}$. The type of external force one has in mind here is essentially the tide-generating force which is generally assumed to derive from a potential (i.e. $\nabla \wedge \bar{\xi} = 0$). $\bar{\xi}$ can then be combined with the pressure and surface elevation gradients.

2.3. TIDAL AND STORM SURGE MODELS

As mentioned in Chapter 1, one of the main problems in hydrodynamic modelling is the specification of boundary conditions on open-sea frontiers (e.g. Nihoul, 1973). The lack of reliable experimental data and the necessity to interpolate from incomplete — often obsolete — observations is a source of errors and misapprehension which can be propagated by the model through the entire system.

Attention has thus been paid essentially so far to oscillating motions for which the open-sea boundary conditions are reasonably easy to approximate, and most models have been concerned with the calculation of tides and storm surges* (e.g. Hansen, 1956; Fischer, 1959; Leendertse, 1967; Kraav, 1969; Heaps, 1969, 1972; Ronday, 1973).

The existing models are essentially depth-averaged models governed by the equations derived in the last section and summarized in Table 2.1. In general, these equations are further simplified by neglecting the surface elevation ζ with respect to the depth h. This approximation limits the applicability of the models to reasonably deep regions of the sea, and the results of the simulation in shallow areas and near the coasts may be significantly affected and differ appreciably from the observations.

The existing models of tides and storm surges also suffer from the lack of sufficiently detailed atmospheric data over the sea. One cannot do much more than estimate the wind stress to the best of one's ability, especially as the value of the drag coefficient C is uncertain and may vary considerably, as pointed out in Chapter 1.

Two other control parameters a and D appear in the equations. They are also difficult to determine with precision, and very often the success of a model depends on a more or less skilful estimate of them.

The approximation of the boundary conditions on open-sea frontiers may finally be responsible for other limitations of the models.

In the case of internal surges for instance it is generally accepted that, if the sea basin is relatively shallow and its open boundary is contiguous with a deep ocean, the surface elevation may be put equal to zero along that boundary for all time. One may fear that this produces erroneous reflections at the open boundary after a time. It seems unlikely that the prediction of sea behaviour after the first transit of the storm surge can be very accurate. For external surges, the open-sea boundary conditions are either measured — in which case they suffer from the general lack of accuracy of experimental data on open-sea boundaries — or approximated by known periodic conditions. This amounts really to replacing a travelling wave by a standing one and again creates the same problems.

By the standards of other fields of simulation, however, it is fair to say that the storm surge and tidal models are at present the most advanced and best prepared for practical applications. To illustrate the art and limitations of these models, it is profitable to discuss briefly one example. It will serve to exemplify the classical approximations and methods and it will give an idea of the typical results such models can achieve.

---

* Limitation to long waves follows from computer speed and memory specifications imposing a relatively coarse grid, which in turn, because of Shannon's sampling theorem, limits resolution to waves with a length equal to or greater than twice the grid-size.

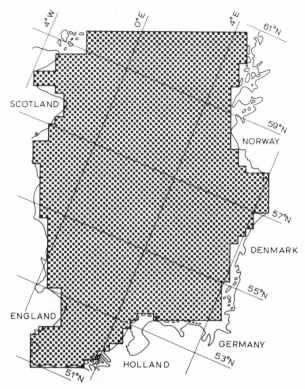

Fig. 2.1. Numerical grid in Ronday's model of the tidal circulation in the North Sea.

Fig. 2.1 shows the numerical grid used by Ronday (1973) to study the circulation induced in the North Sea by external tides originating from the Atlantic Ocean.

As shown by Defant (1961), external tides are largely dominant in coastal seas like the North Sea and the astronomical tide may be neglected. This amounts to setting $\bar{\xi} = 0$ in the governing equations and considering only tidal oscillations induced by incoming tidal waves at the open-sea boundary. Attention is furthermore restricted to the semi-diurnal tide M_2 which is by far the main component in the North Sea.

The discretization scheme proposed by Lax and Wendroff (1960) is used. The horizontal mesh size is $21.7 \cdot 10^3$ m and the associated time step, based on numerical stability criteria, is 1/360 of the M_2 tidal period, i.e. about one minute.

With a typical length scale of some twenty kilometers the eddy diffusivity would be of the order of 10^2 m^2/sec (cf. section 1.7). Following Brett-schneider (1967), Ronday (1973) takes:

$a \sim 10^4$ m^2/sec

As discussed in section 2.2, such a large dispersion coefficient presumably indicates an important contribution of the shear effect in the dispersion of momentum.

Showing that the finally established oscillatory regime is practically independent of the initial conditions, Ronday (1973) assumes zero fluid transport and zero elevation at the initial time.

The boundary conditions along the open-sea frontiers are approximated by simple harmonic functions:

$$H = A \cos{(\omega t + \varphi)} \tag{2.38}$$

where the amplitude A and the phase φ are functions of position. The values of A and φ at the coastal extremities of the open-sea frontiers are given in Table 2.2. The values of A and φ, at boundary points out at sea being unknown, are calculated by linear interpolation.

This illustrates the remark made before about the insufficiency of experimental data and the obligation to elaborate from the existing observations approximate expressions of the boundary conditions. One can see also why wave models are less likely to be affected by the sequels of the boundary approximations as, assuming eq. 2.38 is indeed valid all along the frontier, they influence the coefficients, not the form, of the prescribed boundary data.

Running the simulation program shows that the regime is attained after three tidal periods even for grid points far away from the boundary excitation and situated in shallow areas. This confirms that the solution of the fundamental equations, which corresponds to forced oscillations, is fairly insensitive to the initial conditions. Fig. 2.2 shows the influence of the num-

TABLE 2.2

Values of the amplitude A and phase φ along the open-sea frontiers

	A (m)	φ (degrees)
Straight of Dover		
northern side	2.45	330
southern side	2.70	325
Northern boundary		
western side	0.632	306
eastern side	0.439	241
Baltic Sea		
northern side	0.083	80
southern side	0.108	95

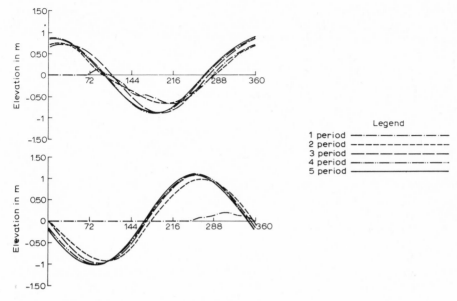

Fig. 2.2. Influence of the number of iterations on the surface elevations at two points respectively at small and large distance from the boundary excitation (after Ronday, 1973).

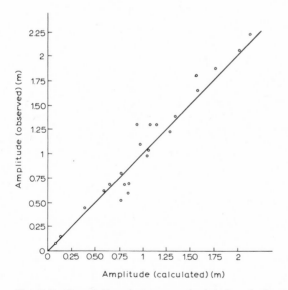

Fig. 2.3. Correlation between calculated and observed amplitudes of the tide M_2 in the North Sea (after Ronday, 1973).

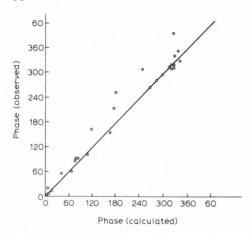

Fig. 2.4. Correlations between calculated and observed values of the phase of the tide M_2 in the North Sea (after Ronday, 1973).

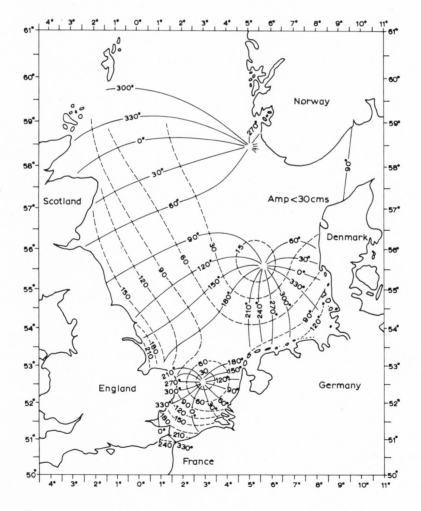

Fig. 2.5. Lines of equal tidal phases and amplitudes in the North Sea according to observations (after Proudman and Doodson, 1924).

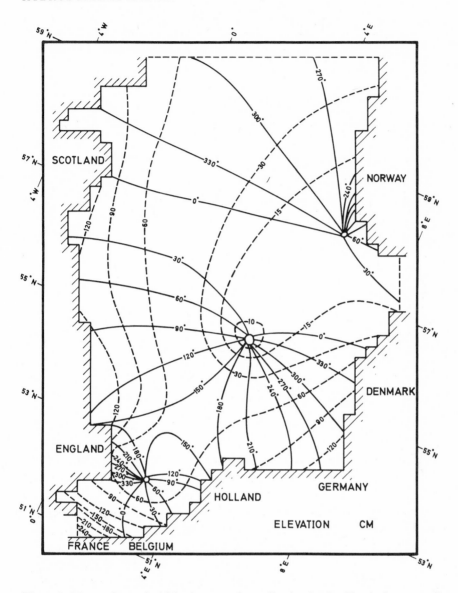

Fig. 2.6. Lines of equal tidal phases and amplitudes in the North Sea according to the mathematical model (after Ronday, 1973).

ber of iterations on the elevation of the sea surface for two points situated respectively at small and large distance from the exciting forces.

Fig. 2.3 and 2.4 show a comparison between observed and calculated values of the tidal amplitude and phase at different coastal stations. There is a generally good agreement except at some points which fall a little beside the ideal line at 45°. These points correspond to places near the English and

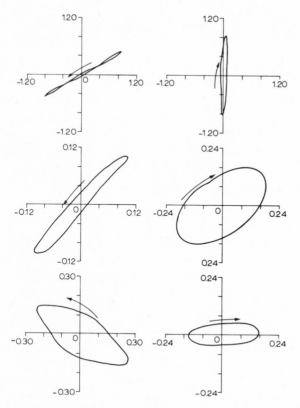

Fig. 2.7. Tidal "ellipses" at some typical points of the North Sea (after Ronday, 1973).

Danish coasts where the depth is small and the bottom very irregular.

The discrepancy here illustrates the general lack of precision of tidal models in shallow areas where non-linear terms are important.

A comparison between observed and calculated lines of equal amplitudes and phases (Fig. 2.5 and 2.6*) also shows an excellent agreement; with again some differences in the *Deutsche Bucht* and along the coast of England where the small water depth is responsible for the effect of non-linear terms.

Ronday (1973) also calculated the tidal currents in the North Sea and compared them with the observations. He found a variance of 5—20% between the calculated and observed currents, the maximum difference occurring at reversing tide when the slightest phase displacement is enough to modify significantly the current pattern because of the marked anisotropy of the tidal "ellipses" (Fig. 2.7).

The differences may be attributed to the approximations of the mathematical and simulation models (discretization, linear interpolation of bound-

* Note that Fig. 2.5 and 2.6 are rotated with respect to one another at some 30°.

ary conditions, etc.), but they cannot be regarded as really significant, considering the relatively rare and indeterminate experimental data.

2.4. MODELS OF RESIDUAL CURRENTS

Frequently, when confronted with a complex system of partial differential equations, one investigates first the existence and characteristics of eventual steady-state solutions. Hence one might find it surprising that the first hydrodynamic models which were developed were concerned with wave motions, tides and storm surges and very little with stationary currents.

There are for that three reasons:

(1) From a numerical point of view, there is little advantage in seeking steady-state solutions because as pointed out by Emmons (1970) "the necessary iterations required to get from a guessed solution to the final steady state (if one exists) are only slightly less work than carrying out the iterations that nature would have gone through to get to the same steady state".

(2) On open-sea boundaries, the data are at present insufficient and, while boundary oscillations are reasonably simple to model, even with sparse experimental data, stationary currents cannot be specified without a detailed knowledge of their distribution along the frontier.

(3) Although steady-state hydrodynamic models have been used successfully to compute wind-driven water circulation in lakes, the possible significance of steady-state solutions for seas and oceans is not, at present, entirely clear.

The third reason is perhaps the most important one. The physical meaning of steady currents is indeed far from obvious. It is generally agreed that water movements which have specially large characteristic times of variation can be approximated by time-independent flows over any reasonable period and the name "residual currents" is broadly accepted to denote such slowly varying motions of the sea.

Different interpretations appear however when specialists undertake to describe residual currents mathematically, simulate them or identify pertinent experimental data. From a mathematical point of view, it is tempting to define the residual currents as the steady flow pattern which is described by the fundamental equations reduced by assuming no time dependence and, consequently, zero derivatives with respect to time.

Experimentalists however often prefer to regard the residual currents as the residuary flow obtained by substracting from the actual fluid motion the (computed) main tidal currents (e.g. Otto, 1970).

Hydrodynamicists have a still different notion. They define the residual currents as mean currents over a time sufficiently long to cover several tidal periods and thus cancel out most of the tidal contributions. The hydrodynamicist's point of view is apparently the more realistic.

One can indeed object to the experimentalist's interpretation that subtracting from the observed values of the actual currents uncertain calculated values of the tidal currents presumably worsens the experimental errors by the additional inexactitudes of the calculus. It is not clear moreover how time-dependent wind currents are eliminated in the process, and the stationary (or almost stationary) character of the result is not obvious.

Simple steady-state solutions of the fundamental equations appear to deserve similar criticisms. Surely steady-state solutions require steady-state forcing and this implies some sort of a long-time average of at least the wind field.

Along the same lines it seems logical to regard the residual current field as the mean field over a time sufficiently long to cancel, to a large extent, transitory wind currents and tidal oscillations.

The residual currents, defined in this way, can only vary very slowly with time and it is reasonable to describe them by steady-state equations. These equations however cannot be obtained directly by dropping the time derivative in the hydrodynamic equations given in Table 2.1. One must first average these equations over time. The average equations will have the same form as far as the linear terms are concerned but they will contain additional contributions from the non-linear terms.

Indeed, if U_0 and H_0 denote the residual (steady) parts of U and H, writing:

$$U = U_0 + U_1 \tag{2.39}$$

$$H = H_0 + \zeta_1 ; \quad H_0 = h + \zeta_0 \tag{2.40}$$

one may assume:

$$(U_1)_0 \sim 0 \tag{2.41}$$

$$(\zeta_1)_0 \sim 0 \tag{2.42}$$

the subscript $_0$ denoting a long-time average.

Thus, averaging a linear term, one eliminates all contributions from U_1 and ζ_1. This is not true, however, for the non-linear terms because mean products of the type $(U_1 U_1)_0$ are not zero.

Neglecting ζ_1 with respect to H_0 for the simplicity of the argument*, one has thus:

$$[\nabla \cdot (H^{-1} UU)]_0 \sim \nabla \cdot (H_0^{-1} U_0 U_0) + \nabla \cdot (H_0^{-1} U_1 U_1)_0 \qquad (2.43)$$

$$[DH^{-2} U\|U\|]_0 \sim DH_0^{-2} U_0 \|U_0\| + DH_0^{-2} U_0 \|U_1\|_0 \qquad (2.44)$$

The residual currents are always considerably smaller than the time-dependent wind and tidal currents. Hence the essential contributions to eqs. 2.43 and 2.44 come from the second terms in the right-hand side and not from the first terms which would have been the only ones to appear if the steady-state approximations had been made directly on the hydrodynamic equations without averaging first.

Neglecting these terms, one finds that the bottom friction is now *linear* in U_0 while the former advection term is independent of U_0 and expresses an apparent forcing stress exerted by the transitory and tidal motions.

Another contribution comes from the quadratic terms in the tidal elevation ζ_1. This contribution has the form $\zeta_1 \nabla g \zeta_1$ and combines with eq. 2.43 in what we might call the "tidal stress".

Inasmuch as the average wind stress τ_s must be determined from observations or atmospheric models, the additional tidal stress, written τ_t in brief, must be estimated from experimental data or tidal and transient circulation models.

In practice, the time average of the dispersions term $a \nabla^2 U$ is small compared with the contributions from the bottom friction, the elevation gradient and the Coriolis effect. If this term is neglected also, the averaged (steady-state) hydrodynamic equations can be written (taking $\bar{\xi}_0 \sim 0$, in the present context):

$$f e_3 \wedge U_0 = -H_0 \nabla \left(\frac{p_a}{\rho} + g\zeta_0\right) - KH_0^{-1} U_0 + \vartheta \qquad (2.45)$$

where:

$$\vartheta = \rho^{-1}\tau ; \qquad \tau = \tau_s + \tau_t \qquad (2.46)$$

---

* The continuity eq. 2.20 gives:

$$\frac{\partial \zeta_1}{\partial t} + \nabla \cdot U_1 = 0 , \quad \text{i.e.} \quad \zeta_1 \sim 0\left(\frac{H\bar{u}_1}{c}\right)$$

where c is the ratio of the typical length and time scales of the time-dependent components U_1 and ζ_1 (the phase velocity for tidal motions). In general, $\bar{u}_1 \ll c$ and $\zeta_1 \ll H$. Keeping ζ_1 in the expression of H would however not modify the argument presented here but it would unnecessarily cumber the presentation.

and where:

$$K = D \, \|\bar{u}_1\|_0 \tag{2.47}$$

is a new friction coefficient.

In addition, averaging the continuity equations one obtains:

$$\nabla \cdot U_0 = 0 \tag{2.48}$$

Hence, the two components of the vector U_0 can be derived from a stream function Ψ such that:

$$U_{0,1} = -\frac{\partial \Psi}{\partial x_2} \tag{2.49}$$

$$U_{0,2} = \frac{\partial \Psi}{\partial x_1} \tag{2.50}$$

(Eq. 2.48 is then identically satisfied.)

Dividing eq. 2.45 by H_0 and taking the curl to eliminate the surface elevation, one obtains, substituting 2.49 and 2.50:

$$K \, \nabla^2 \Psi - \frac{\partial \Psi}{\partial x_1} \left(f \frac{\partial H_0}{\partial x_2} + \frac{2K}{H_0} \frac{\partial H_0}{\partial x_1} \right) + \frac{\partial \Psi}{\partial x_2} \left(f \frac{\partial H_0}{\partial x_1} - \frac{2K}{H_0} \frac{\partial H_0}{\partial x_2} \right)$$

$$= H_0 \omega_z + \frac{\partial H_0}{\partial x_2} \theta_1 - \frac{\partial H_0}{\partial x_1} \theta_2 \tag{2.51}$$

where:

$$\omega_z = (\nabla \wedge \boldsymbol{\vartheta})_z \tag{2.52}$$

The problem thus reduces to a single partial differential equation for the stream function Ψ. In the particular case of constant depth and irrotational stress $\boldsymbol{\tau}$, this equation reduces to the well-known Laplace equation:

$$\nabla^2 \Psi = 0 \tag{2.53}$$

In general, however, there is an important influence of the bottom topography on the residual currents pattern.

This is very clearly illustrated by the following example. In his study of the residual currents in the North Sea, Ronday (1972) discussed in detail the circulation during the month of January. The model is shown in Fig. 2.8. Eq. 2.51 was solved by Liebman's method of forced relaxation (e.g. Murty and Taylor, 1970). Boundary conditions for the stream function Ψ were inferred from observations on the different sections of the frontier (cf.

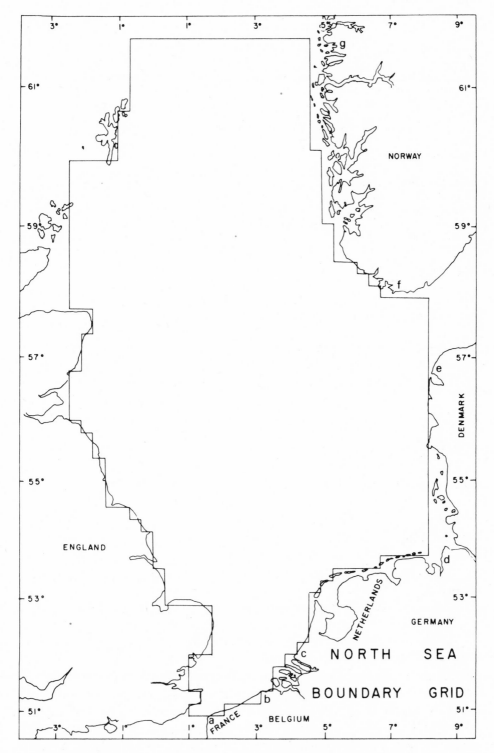

Fig. 2.8. Contours of numerical grid for the model of residual circulation in the North Sea, indicating positions of the boundary where different boundary conditions prevail (after Ronday, 1972).

TABLE 2.3

Boundary conditions on the stream function Ψ for the model of residual circulation in the North Sea (according to Ronday, 1972)
(Ψ is given in m^3/sec)

---

1. Along the coast of Great Britain from the northwest frontier to Dover:
   $\Psi = 0$
2. Along the continental coast from Calais (a) to Flushing (b):
   $\Psi = 2.40 \cdot 10^5$
3. From Hoek van Holland (c) to Cuxhaven (d):
   $\Psi = 2.42 \cdot 10^5$
4. From Cuxhaven to the southern frontier of the Skagerrak (e):
   $\Psi = 2.43 \cdot 10^5$
5. Along the Norwegian coast (f) to the northeast frontier of the model (g):
   $\Psi = 2.70 \cdot 10^5$
6. Across the Straits of Dover from $\Psi = 0$ to $\Psi = 2.40 \cdot 10^5$.
7. Across the northern frontier:
   (a) from $\Psi = 2.7 \cdot 10^5$ at the Norwegian coast to $\Psi = -2918$ at depth $H_0 = 225$ m
   (b) from $\Psi = -2918$ at $H_0 = 225$ m to $\Psi = 0$ at the northwest corner of the grid.

Fig. 2.8). Along the open-sea boundaries linear interpolations were used. The boundary conditions are summarized in Table 2.3.

From experimental data, Ronday (1972) also estimates:

$$\omega_z \sim 0 \tag{2.54}$$

$$\theta_1 \sim \theta_2 \sim 7 \cdot 10^{-5} \, m^2/sec^2 \tag{2.55}$$

The calculated stream-lines are shown in Fig. 2.9. The general pattern agrees fairly well with the qualitative picture proposed by Laevastu (1963) on the basis of the observations (Fig. 2.10).

The most important characteristics of the circulation are well depicted:

(1) *Inflow of North Atlantic water.* The area between the lines $\Psi = 0$ and $\Psi = -10^{-5}$, extending from the northern frontier to the 56° parallel corresponds to North Atlantic water (region *1* on the map proposed by Laevastu).

(2) *Inflow of Baltic water.* The area between the steam-line $\Psi = -3 \cdot 10^5$ and the Norwegian coast corresponds to region *3* of Laevastu.

(3) *Inflow from the Straits of Dover.* The stream-line $\Psi = 10^5$ separates the North Atlantic water and the water coming from the Straits of Dover (region *2* and *6* of Laevastu).

(4) *Reversal of the North Atlantic flow at the southwest of the Norwegian coast.* The circulation around the Dogger Bank is correctly reproduced.

Fig. 2.11 shows the calculated stream-lines when a constant depth is assumed. One can see that the general pattern is strongly modified and no longer agrees with the observations. In particular the penetration of the

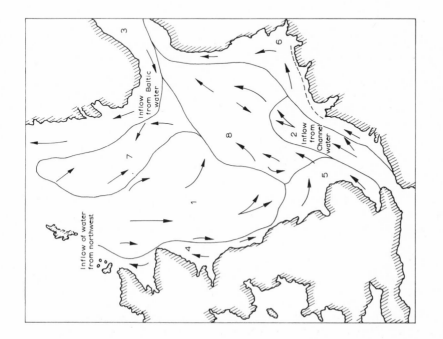

Fig. 2.9. Residual stream-lines ψ = constant in the North Sea (after Ronday, 1972). The actual values of ψ are 10^4 times the indicated figures.

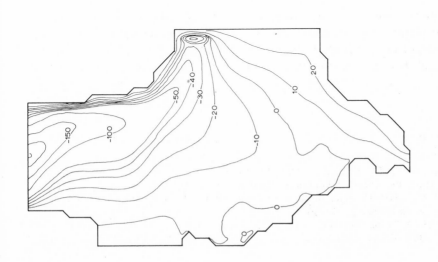

Fig. 2.10. Water masses in the North Sea according to Laevastu (1963).

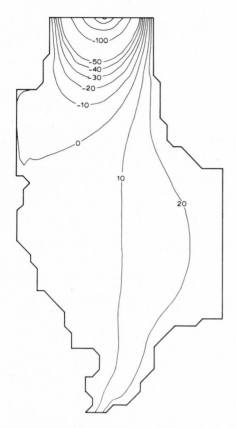

Fig. 2.11. Residual stream-lines Ψ = constant in the North Sea calculated, assuming constant depth (after Ronday, 1972). The actual values of Ψ are 10^4 times the indicated figures.

North Atlantic water is reduced, the flow reverses higher North and the circulation in the central part of the North Sea is completely different from what one might infer from the observed distribution of water masses.

This illustrates the cogent influence of bottom topography on the residual currents.

REFERENCES

Brettschneider, G., 1967. *Mitt. Inst. Meereskd. Univ. Hamburg*, 7.
Defant, A., 1961. *Physical Oceanography*. Pergamon Press, Oxford, 745 pp.
Emmons, H.W., 1970. Critique of numerical modelling of fluid-mechanics phenomena. *Ann. Rev. Fluid Mech.*, 2: 15—36.
Fischer, G., 1959. *Tellus*, 11: 611.
Hansen, W., 1956. *Tellus*, 8: 287.

Heaps, N.S., 1969. *Philos. Trans. R. Soc. Lond.*, 265A: 93.

Heaps, N.S., 1972. In: *Proc. Liège Colloq. Ocean Hydrodynamics, 3rd, Liège, 1971. Mém. Soc. R. Sci. Liège*, 2: 143.

Kraav, V.K., 1969. *Akad. Nauk S.S.S.R.*, 9: 333—341.

Laevastu, T., 1963. *Serial Atlas of the Marine Environment.* Am. Geogr. Soc., Washington, D.C. (Folio 4, 2 charts.)

Lax, P.D. and Wendroff, B., 1960. *Commun. Pure Appl. Math.*, 13: 217.

Leendertse, J.J., 1967. *Aspects of a Computational Model for Long-Period Water-Wave Propagation.* Thesis, Technische Hogeschool, Delft.

Murty, T.S. and Taylor, J.D., 1970. *J. Ocean Soc. Jap.*, 26 (4): 15.

Nihoul, J.C.J., 1973. In: E. Goldberg (Editor), *North Sea Science Book.* M.I.T. University Press, Cambridge, Mass., pp. 43—57.

Otto, L., 1970. The mean residual transport pattern in the southern North Sea. *Int. Counc. Explor. Sea*, CM 1970, C21.

Proudman, J. and Doodson, A.T., 1924. *Philos. Trans. R. Soc. Lond.*, 224A: 185.

Ronday, F.C., 1972. Modèle mathématique pour l'étude de la circulation résiduelle dans la mer du Nord. *Marine Science Branch, Manuscript Report Series*, 27 (Ottawa).

Ronday, F.C., 1973. Modèle mathématique pour l'étude de la circulation due a la marée en mer du Nord. *Marine Science Branch, Manuscript Report Series*, 29 (Ottawa).

CHAPTER 3

PASSIVE DISPERSION MODELS

Jacques C.J. Nihoul

3.1. INTRODUCTION

One of the main objectives of marine models is the prediction of the distribution in space and time of temperature, nutrients, pollutants, etc.

As shown in Chapter 1 (cf. Fig. 1.8), the evolution of a marine constituent results from the advection by the currents, the migration (in particular the sedimentation), the dispersion by molecular diffusion, turbulence and in general all small-scale motions which contribute to the agitation of the sea, and the chemical, biochemical or ecological interactions.

The cumulative effect of these processes on the space and time variations of the concentration r_α of a given constituent α is expressed in mathematical form by eq. 1.19.

Advection and dispersion are governed by the sea dynamics. There is thus a coupling between the state variable r_α and the mechanical variables describing the system's hydrodynamics. In general also, the interactions introduce a coupling between r_α and the concentrations of other constituents.

If the hydrodynamics is known experimentally or given by a preliminary model, one can calculate the velocity field and the dispersion coefficients and substitute them in eq. 1.19. If, furthermore, the interactions with other constituents have only little influence on the evolution and, in the first instance, can be neglected, eq. 1.19 can be solved independently of the other evolution equations of the model.

A non-interacting constituent which can be described in this simple way is called a *passive constituent* and the model which is subsequently reduced to a single evolution equation with given mechanical coefficients of advection and dispersion is generally called a *passive dispersion model*. It is obviously important in marine modelling to master the techniques of passive dispersion models.

In addition to adequately describing the evolution of some — effectively passive — constituents, such models always provide valuable estimates of the distribution of non-passive constituents by, at least, appraising their possible transport by the sea motions.

Moreover, in many cases, the interaction processes are not yet entirely understood and cannot be formulated with sufficient accuracy to be used in a reliable simulation model. It is then preferable to base one's predictions on the estimates of a passive dispersion model until new experimental and theoretical data are available and the interaction terms can be determined with the required precision.

In any case, if the interactions play a significant role, they can be taken into account in a passive dispersion model if one assumes that their net result is a production or destruction of constituent α which can be expressed as a function of r_α alone.

Indeed, the condition for eq. 1.19 to be independent of the other evolution equations is that the interaction term I_α be zero or a function of r_α only. This is the case, for instance, with a radioactive substance which decays in time at a rate proportional to its concentration.

On the model of radioactive decay, the resulting effect of all interactions is often approximated by a linear production or destruction term of the form:

$$I_\alpha = kr_\alpha \tag{3.1}$$

With this approximation, passive dispersion models can be used to describe the general features of the space and time variations of even non-passive constituents.

This is again an example of the general policy which consists in extracting, from the exhaustive multi-purpose mathematical model one tries to elaborate, the elements of less ambitious but more practical sub-models to accelerate simulation, prediction and diagnosis.

3.2. TURBULENT DISPERSION

The last term in the right-hand side of eq. 1.19 represents the dispersion of the marine constituents by the agitation of the sea.

The mathematical formulation of the dispersion effect has been discussed in Chapter 1, and the applicability of diffusion formulas has been critically examined. One has shown in particular that the appropriate eddy diffusion coefficients were much larger than the molecular diffusivity, that they could be functions of time and space with different values in different directions.

In passive dispersion models, one generally assumes that horizontal dispersion is reasonably homogeneous and isotropic and can be described with one dispersion coefficient (e.g., κ).

Introducing another coefficient of dispersion λ in the vertical direction, one writes:

$$\mathcal{D}_\alpha = \nabla_h \cdot (\kappa \nabla_h r_\alpha) + \frac{\partial}{\partial x_3} \left(\lambda \frac{\partial r_\alpha}{\partial x_3} \right) \tag{3.2}$$

where:

$$\nabla_h \equiv e_1 \frac{\partial}{\partial x_1} + e_2 \frac{\partial}{\partial x_2} \tag{3.3}$$

The disparity between the horizontal and vertical eddy diffusivities is a natural consequence of the different scales of horizontal and vertical motions. The stable stratification of the water column also affects the vertical diffusion and contributes to reinforce the difference between the two diffusivities.

The vertical stratification is usually described in terms of the mean buoyancy:

$$a = g \frac{r - r_0}{r_0} \tag{3.4}$$

where, in accordance with the notations introduced in Chapter 1 (Table 1.1), g is the acceleration of gravity, r the mean water density and r_0 a reference value of r.

The x_3-axis pointing upwards, one defines:
(1) the Brunt-Väisälä frequency:

$$N = \left(-\frac{\partial a}{\partial x_3} \right)^{1/2} \tag{3.5}$$

(2) the Kelvin-Helmholtz frequency:

$$M = \left\| \frac{\partial u}{\partial x_3} \right\| \tag{3.6}$$

(3) the Richardson number:

$$Ri = \frac{N^2}{M^2} \tag{3.7}$$

Several essential aspects of the dynamics of a stably stratified fluid result from the local competition of two important effects: release of energy from the mean field to the turbulent motion and removal of energy from the turbulent field by the buoyancy forces. The Richardson number is a measure of the relative efficiency of the two processes. One finds that laminar motion is stable for $Ri > 0.25$. Once turbulence is generated, however, it can persist for higher values of the Richardson number (Stewart, 1959); the Richardson number then appreciates the ability of the turbulent motion to support itself.

In shallow waters, the velocity shear M is always very large and active turbulence is produced, resulting in an often complete mixing of the water column. In the open sea, the same type of energetic turbulence is observed in the upper layers where it is produced by the wind. Below the wind-mixed layer, strong temperature and density gradients persist in the thermocline, and observations indicate that the turbulence there is highly intermittent and sporadic (Grant et al., 1968).

Several investigations (Stommel and Fedorov, 1967; Cooper and Stommel, 1968; Woods, 1968a,b) have revealed that the thermocline is divided into a series of fairly uniform "layers" separated by thin interfacial regions ("sheets") of large temperature and density gradients where extreme shears can also exist.

Internal waves can propagate along the sheets. Unstable waves eventually grow, sharpen, turn over and break, emitting patches of turbulence which maintain the intermittent turbulence of the layers around.

In deep waters, very little is known about possible turbulent motions. The velocity shear associated with the general — almost geostrophic — flow is apparently too small to produce instability. Weak shears, created by internal tides, for instance (Munk, 1966), could however be sufficient to overcome the moderate stratification and generate turbulence.

In any case, the rate of turbulent energy transfer ϵ seems to decrease from the naviface (where especially high values are probably due to breaking surface waves) to the deep sea. Typical values are given in Table 3.1.

As explained in Chapter 1, the cascade process, by which breaking eddies transfer energy to smaller ones, produces a wide spectrum of turbulent eddies between the large-scale eddies developing from the major ocean currents and the small-scale ($\lesssim 10^{-2}$ m) eddies where viscous dissipation takes place. Although *complete* isotropy may not be expected for scales larger than a few

TABLE 3.1

Rate of turbulent energy transfer ϵ at different depths

Depth (m)	ϵ (m^2/sec^3)	Reference
1 or 2 m below naviface	$3 \cdot 10^{-5}$	Stewart and Grant (1962)
mixed layer (~ 30 m)	$3 \cdot 10^{-6}$	Grant et al. (1968)
thermocline	$3 \cdot 10^{-8}$	Woods (1968b)
deep water (~ 500 m)	$3 \cdot 10^{-9}$	Takenouti et al. (1962)
abyssal depth ($10^3 \rightarrow 4 \cdot 10^3$ m)	$3 \cdot 10^{-10}$	Munk (1966)

meters, *horizontal* homogeneity and isotropy may be assumed up to much larger scales, and Ozmidov (1965) has argued that Kolmogorov's theory could be applied to horizontal turbulence provided the inputs of energy at intermediate scales were taken into account and a different value of the rate of energy transfer ϵ was used in each of three distinct bands of wave numbers (cf. section 1.7).

Thus, according to Ozmidov the horizontal eddy diffusivity would be given, in terms of the scale ℓ, by:

$$\kappa \sim \epsilon^{1/3} \ell^{4/3} \tag{3.8}$$

similar to eq. 1.59 for the eddy viscosity.

Although it is generally agreed that, as a result of the wide spectrum of horizontal motions, the horizontal eddy diffusivity must increase with the scale of mixing considered, there has been so far no really convincing evidence that it behaves exactly as the 4/3 power of ℓ.

According to Joseph and Sendner (1958), the horizontal eddy diffusivity should be proportional to the length scale of the mixing process. Introducing a *diffusion velocity* P, they take:

$$\kappa \sim P\ell \tag{3.9}$$

where the diffusion velocity — which is related to the root-mean-square value of the velocity fluctuations — is assumed to be constant $(P \sim 10^{-2}$ m/sec)*.

Okubo (1968), reviewing data from a large number of experiments, derived the relation:

$$\kappa \sim 0.0103 \, \ell^{1.15} \tag{3.10}$$

The exponent 1.15 is intermediate between the value 1 given by Joseph and Sendner and the value 4/3 advocated by Ozmidov.

Results from a diffusion experiment in the North Sea reported by Joseph et al. (1964) suggest on the contrary that the exponent is less than 1 and actually decreases to 0 with increasing scale, the eddy diffusivity tending to a constant value. A similar behaviour was apparently observed by Talbot (1970).

As pointed out by Okubo (1968), the experimental data cannot definitely prove or rule out any theory. One must remember in particular that coefficients like ϵ or P are functions of time and may vary appreciably during a series of observations. Such variations produce a scattering of the experimental data through which, in most cases, curves like 3.8, 3.9 or 3.10 can be drawn with an equally good fit.

---

* Observations reported by Joseph and Sendner later (1962) indicated however a weak dependence of P on ℓ.

The three curves are, however, very close and they all provide good estimates of the horizontal eddy diffusivity when the control parameters (ϵ, P,
etc.) are properly adjusted.

One may see in this agreement a justification of the concept — everywhere
subjacent — of homogeneous horizontal turbulence where each scale of motion is basically animated by energy transferred from larger scales through a
descending cascade of eddies.

In contrast, vertical turbulence is essentially inhomogeneous and governed
by the exchanges of energy with the mean fields. Kinetic energy is extracted
from the mean velocity field and, in stably stratified water, partly recovered
as potential energy by the buoyancy forces.

It is not surprising then that most of the formulas proposed for the
vertical eddy diffusivity were inspired by shear flow turbulence (as in pipes
and channels) rather than homogeneous turbulence.

In general the vertical diffusivity λ is assumed proportional to the mean
shear M (eq. 3.6). In neutrally stratified water, one writes:

$$\lambda \sim \ell_m^2 \, M \tag{3.11}$$

introducing the *mixing length* ℓ_m which one usually determines from observations and sideways theoretical reflections.

The mixing length may be a function of time even if, for simplicity, it is
often assumed constant. For instance, Nihoul (1973a) suggested that, in the
upper mixed layer of the sea, ℓ_m is proportional to the depth h of the layer.
(The factor of proportionality inferred from a large number of experiments
was found to be of order 10^{-1}.) With this assumption he was able to derive
a model of the turbulent mixing and entrainment which predicted mean
velocity and buoyancy profiles in the turbulent region in good agreement
with the observations.

Here, the mixing length is a function of time because, as long as it is
fostered by the wind, the turbulence erodes and entrains the laminar fluid
below and the depth of the mixed layer increases with time.

In highly turbulent estuaries or shallow seas or in the upper mixed layer of
the ocean, the Richardson number is small and one may neglect the effect of
stratification on the vertical diffusivity.

In stably stratified water however, vertical turbulence can be partly inhib-
ited and the vertical diffusivity accordingly reduced. In general, this effect is
accounted for by allowing the diffusivity to depend on the Richardson number.

For instance, λ_0 denoting the eddy diffusivity in the neutral case, Mamayev (1958) proposes:

$$\lambda \sim \lambda_0 \, e^{-0.8 \, Ri} \tag{3.12}$$

TABLE 3.2

Vertical eddy diffusivity λ at different depths

Depth	λ (m²/sec)	Reference
upper mixed layer	10^{-2}	Nihoul (1973a)
thermocline	10^{-6}	Kullenberg (1970)
deep layers below the thermocline	10^{-4}	Okubo (1971)

while Munk and Anderson (1948) write:

$$\lambda \sim \lambda_0 \, (1 + 3.33 \, Ri)^{-3/2} \tag{3.13}$$

The reduction of the vertical diffusivity is more severe in the thermocline (or any layer of strong stability) where the Richardson number is larger. Typical values of λ are given in Table 3.2.

3.3. THE EQUATION OF PASSIVE DISPERSION

If one assumes that the interaction term I_α in eq. 1.19 is either zero or a linear function of r_α alone and if the eddy diffusion approximation 3.2 is made for $\mathcal{D}_\alpha$, eq. 1.19 reduces to the linear partial differential equation:

$$\frac{\partial r_\alpha}{\partial t} + \nabla \cdot r_\alpha \boldsymbol{u} = Q_\alpha + k r_\alpha - \nabla \cdot r_\alpha \boldsymbol{\sigma}_\alpha + \nabla_h \cdot (\kappa \nabla_h r_\alpha) + \frac{\partial}{\partial x_3} \left(\lambda \frac{\partial r_\alpha}{\partial x_3} \right) \tag{3.14}$$

It is convenient to eliminate the term $k r_\alpha$ from eq. 3.14. Let:

$$r_\alpha = c(\boldsymbol{x}, t) \, e^{kt} \tag{3.15}$$

Substituting in eq. 3.14 and dividing by e^{kt} one gets, dropping the subscript α which is now superfluous as one considers only one constituent:

$$\frac{\partial c}{\partial t} + \nabla \cdot c \boldsymbol{u} = S - \nabla \cdot c \boldsymbol{\sigma} + \nabla_h \cdot (\kappa \nabla_h c) + \frac{\partial}{\partial x_3} \left(\lambda \frac{\partial c}{\partial x_3} \right) \tag{3.16}$$

where

$$S = Q_\alpha \, e^{-kt} \tag{3.17}$$

is a given function of x and t representing external sources and sinks.

As explained before, the vertical and horizontal motions are characterized by completely different scales. It is profitable to emphasize the distinction between horizontal and vertical advection and migration as one did for the dispersion.

Introducing horizontal advection and migration velocities:

$$u_h = u_1 \, e_1 + u_2 \, e_2 \tag{3.18}$$

$$\sigma_h = \sigma_1 \, e_1 + \sigma_2 \, e_2 \tag{3.19}$$

eq. 3.16 can be written:

$$\frac{\partial c}{\partial t} + \nabla_h \cdot (c u_h) + \frac{\partial}{\partial x_3} (c u_3)$$

$$\tag{3.20}$$

$$= S - \nabla_h \cdot (c \sigma_h) - \frac{\partial}{\partial x_3} (c \sigma_3) + \nabla_h \cdot (\kappa \nabla_h c) + \frac{\partial}{\partial x_3} \left(\lambda \frac{\partial c}{\partial x_3} \right)$$

The roles of the different terms of eq. 3.20 are easily identified. The variation of the concentration c with time results from the combined effects of

(1) horizontal advection: $\nabla_h \cdot (c u_h)$,

(2) vertical advection: $\dfrac{\partial}{\partial x_3} (c u_3)$,

(3) horizontal migration: $- \nabla_h \cdot (c \sigma_h)$,

(4) vertical migration: $- \dfrac{\partial}{\partial x_3} (c \sigma_3)$,

(5) horizontal diffusion: $\nabla_h \cdot (\kappa \nabla_h c)$,

(6) vertical diffusion: $\dfrac{\partial}{\partial x_3} \left(\lambda \dfrac{\partial c}{\partial x_3} \right)$,

(7) local production or
     destruction by external
     sources and sinks: S .

Vertical advection is usually small compared with horizontal advection.

Horizontal migration is the attribute of fish. This effect is difficult to assess and in most cases only sparse statistical data are available. This perhaps indicates that deterministic models of the fishery dynamics are at present unrealistic, and it may be wise at this stage to exclude fish populations from the state variables of the marine models which are advocated in this volume. In that case, one can neglect horizontal migration.

Vertical migration is essentially associated with sedimentation and ascension of lighter fluid and gases.

Since the vertical diffusivity is generally several orders of magnitude smaller than the horizontal diffusivity one might think that vertical diffusion can be neglected. In fact, it is necessary to distinguish here between constituents which are confined to relatively thin homogeneous layers near the surface (or inside the fluid at discharge level) and essentially spread in a horizontal plane, and constituents which are distributed over the whole water column or the mixed layer. In the first case, vertical variations are very small and the dispersion is basically a two-dimensional horizontal problem. In the second case, the vertical concentration profiles can make the derivatives with respect to x_3 quite important. Large vertical gradients associated with small vertical length scales can compensate the small vertical diffusivity and vertical diffusion, then, often dominates horizontal diffusion.

If one considers, for instance, a turbulent layer of depth h and denotes by u_* the typical value of the shear velocity, one has (e.g. Nihoul, 1973a):

$$\nabla_h \cdot (\kappa \nabla_h c) \sim O\left(\frac{Pc}{\ell}\right) \sim O\left(\frac{u_* c}{\ell}\right)$$

$$\frac{\partial}{\partial x_3}\left(\lambda \frac{\partial c}{\partial x_3}\right) \sim O\left(\frac{\ell_m u_* c}{h^2}\right) \sim O\left(\frac{\alpha u_* c}{h}\right)$$

where

$$\alpha \sim O(10^{-1})$$

Hence the ratio of the horizontal and vertical diffusion terms is of the order:

$$\frac{\text{horizontal diffusion}}{\text{vertical diffusion}} \sim O\left(\frac{10 h}{\ell}\right)$$

Hence the vertical diffusion dominates provided the horizontal length scale is at least ten times larger than the vertical length scale.

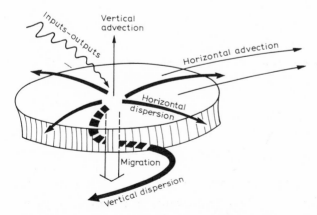

Fig. 3.1. Schematic diagram of the different effects affecting the evolution of a passive constituent.

This is a very frequent situation. In many cases, important variations are observed in the vertical direction while quasi-horizontal homogeneity prevails over a substantially large region. One has then $\ell \gg h$.

The different effects commanding the evolution of a passive constituent are summarized in Fig. 3.1.

3.4. HORIZONTAL DISPERSION

One of the most important problems of passive dispersion is the diffusion of a contaminant, released at a given point of the marine region and subsequently dispersed by sea motions around the point of release. This problem is associated with the accidental or organized dumping of pollutants, the discharge of sewage and used industrial waters, etc.

In many cases, the diffusion may be considered to be two-dimensional, the contaminant being confined to a thin homogeneous layer at discharge level and essentially spreading in a horizontal plane. One can then neglect the vertical components of advection, migration and diffusion. If furthermore the horizontal velocity field may be assumed reasonably uniform over the area of dispersion, the horizontal advection reduces to a simple translation.

Thus, if the x_1-axis is taken in the direction of the translation and if u_0 is the translation velocity, eq. 3.20 can be written in the simple form:

$$\frac{\partial c}{\partial t} + u_0 \frac{\partial c}{\partial x_1} = S + \nabla_h \cdot (\kappa \nabla_h c) \tag{3.21}$$

It is convenient to change to axes moving with the horizontal velocity $u_0 e_1$. The new coordinates x_1^*, x_2^* are given by:

$$x_1^* = x_1 - u_0 t \tag{3.22}$$

$$x_2^* = x_2 \tag{3.23}$$

Hence, the time derivative becomes:

$$\frac{\partial c}{\partial t} - u_0 \frac{\partial c}{\partial x_1^*}$$

and substituting in eq. 3.21 one gets:

$$\frac{\partial c}{\partial t} = S + \nabla_h \cdot (\kappa \nabla_h c) \tag{3.24}$$

Of particular interest is the case of an instantaneous release at a given time. If the time of release is taken as the initial time, the inputs into the system, from the outside, are zero, except at $t = 0$. They thus appear in the initial conditions and, in eq. 3.24, $S = 0$.

Assuming horizontal isotropy in the axes moving with the center of the patch, in polar coordinates r, θ, the concentration c is a function of r and t only.

The eddy diffusivity κ is also, in general, a function of r because it depends on the scale of mixing ($\ell \sim r$). The formulas derived in section 3.2 are all of the form:

$$\kappa = \alpha r^q \qquad . \tag{3.25}$$

The particular values:

$$\alpha = C^{nt}, \quad q = 0$$

correspond to the case of constant diffusivity. In general however, α is a function of time. The formulas proposed by Ozmidov and Joseph are particular cases of eq. 3.25 given respectively by

(1) Ozmidov:

$$\alpha = \epsilon^{1/3} \tag{3.26}$$

$$q = \tfrac{4}{3} \tag{3.27}$$

(2) Joseph:

$$\alpha = P \tag{3.28}$$

$$q = 1 \tag{3.29}$$

Using 3.25, eq. 3.24 reduces to:

$$\frac{\partial c}{\partial t} = \frac{1}{r} \frac{\partial}{\partial r} \left(\alpha r^{q+1} \frac{\partial c}{\partial r} \right) \tag{3.30}$$

The conditions of validity of this equation are obviously related to the three hypotheses made: (1) vertical homogeneity in a thin layer where the contaminant is confined, (2) horizontal homogeneity over some region embodying the dispersion patch, and (3) horizontal isotropy in axes moving with the center of the patch.

Although the first two hypotheses constitute a valuable first approximation to estimate the speed and the extent of the dispersion, a weak vertical or horizontal shear may modify the dispersion pattern in a non-negligible way. This effect will be examined later.

The assumption of horizontal isotropy may not seem very realistic if one examines the shape of the patches observed in dispersion experiments. Indeed, while all solutions of eq. 3.30 have a radial symmetry, dispersion patches often show irregular elongated shapes which do not seem to be compatible with the assumed symmetry.

One must remember however that, to eliminate small-scale fluctuations, an average has been performed. Thus c must be regarded either as an average over a large number of similar situations ("ensemble" average) or, practical- ly, as an average over a period of time sufficiently large to smooth out the fluctuations and sufficiently small, as compared with the dispersion time scale, to leave the main dispersion process unaffected.

One should not be surprised then if the observation of instantaneous concentrations in any one particular experiment reveals an irregular pattern which the average equation cannot predict. Only by superposing several patterns of dispersion from many similar experiments, or by appropriately averaging the observed data, one may expect to approach the radially symmetrical distribution predicted by the theory.

Nevertheless the theoretical symmetrical solutions allow simple estimates of the extent and rapidity of the dispersion. They can be used to calculate the effective eddy diffusivity from field observations and experiments with tracers and it is interesting to discuss them briefly here.

In general, solutions of eq. 3.30 are sought in the form:

$$c = \gamma(t) \, e^{-\sigma(t) r^m} \tag{3.31}$$

Substitution in 3.30 gives:

$$\frac{d\gamma}{dt} - \gamma \frac{d\sigma}{dt} r^m = m^2 \sigma^2 \alpha \gamma r^{2m+q-2} - (q+m) \alpha \gamma \sigma \, m r^{q+m-2} \tag{3.32}$$

The existence of a similarity solution of the type 3.31 depends on the possibility of choosing m in such a way that the same powers of r appear in the two members. It is readily seen that the only sensible choice is ($q \neq 2$):

$$q + m - 2 = 0 \tag{3.33}$$

$$2m + q - 2 = m \tag{3.34}$$

and

$$\frac{d\gamma}{dt} = -2(2-q) \, \alpha \gamma \sigma \tag{3.35}$$

$$\frac{d\sigma}{dt} = -(2-q)^2 \, \alpha \sigma^2 \tag{3.36}$$

Solving eq. 3.36 for σ, substituting in 3.35 and integrating, one gets:

$$\sigma^{-1} = (2-q)^2 \int_0^t \alpha \, dt \tag{3.37}$$

$$\gamma = A \, \sigma^{\frac{2}{2-q}} \tag{3.38}$$

where A is a constant of integration.

Hence:

$$c = A \, \sigma^{\frac{2}{2-q}} \, e^{-\sigma r^{2-q}} \tag{3.39}$$

When t tends to zero, σ tends to infinity and c tends to zero everywhere except at $r = 0$. The similarity solution 3.39 is thus the appropriate solution for the problem of an instantaneous point release at $r = 0$, $t = 0$.

If C is the total amount released per unit depth — one remembers here that the contaminant is actually uniformly distributed over a thin layer —, the constant of integration A can be determined in terms of C either by imposing the initial condition or by expressing the conservation of mass:

$$\int_0^{2\pi} d\theta \int_0^\infty cr \, dr = 2\pi \int_0^\infty cr \, dr = C \tag{3.40}$$

Using 3.39, the left-hand side of eq. 3.40 can be written:

$$2\pi A \int_0^\infty \sigma^{\frac{2}{2-q}} \, e^{-\sigma r^{2-q}} \, r \, dr = \frac{2\pi A}{2-q} \int_0^\infty e^{-\xi} \, \xi^{\frac{2}{2-q}-1} \, d\xi \tag{3.41}$$

$$= \frac{2\pi A}{2-q} \, \Gamma\left(\frac{2}{2-q}\right)$$

where Γ is the gamma function.

Hence:

$$A = \frac{(2-q)C}{2\pi \, \Gamma\left(\dfrac{2}{2-q}\right)} \tag{3.42}$$

and

$$c = \frac{(2-q)C}{2\pi \, \Gamma\left(\dfrac{2}{2-q}\right)} \, \sigma^{\frac{2}{2-q}} \, e^{-\sigma r^{2-q}} \tag{3.43}$$

Special forms of eq. 3.43 have been used by several authors assuming constant α (in eq. 3.37) and q either 0, 1 or 4/3. With these assumptions σ is proportional to t^{-1} and the following simple formulas are obtained:

(1) $\alpha = C^{nt}$, $q = 0$; constant eddy diffusivity:

$$c = \frac{C}{4\pi\alpha t} \, e^{-\frac{r^2}{4\alpha t}} \tag{3.44}$$

(2) $\alpha = C^{nt}$, $q = 1$; Joseph's formula for the eddy diffusivity:

$$c = \frac{C}{2\pi\alpha^2 t^2} \, e^{-\frac{r}{\alpha t}} \tag{3.45}$$

(3) $\alpha = C^{nt}$, $q = 4/3$; Ozmidov's formula for the eddy diffusivity:

$$c = \frac{C}{6\pi t^3} \left(\frac{9}{4\alpha}\right)^3 e^{-\frac{9r^{2/3}}{4\alpha t}}$$

(3.46)

Eq. 3.43 shows that the concentration at the center of the patch ($r = 0$) is given by:

$$c_0 = \frac{(2-q)\,C}{2\pi\,\Gamma\left(\dfrac{2}{2-q}\right)}\,\sigma^{\frac{2}{2-q}}$$

(3.47)

Hence, combining 3.43 and 3.47 and setting:

$$\eta = \ln\ln\frac{c_0}{c}$$

(3.48)

$$\xi = \ln\frac{r}{r_1}$$

(3.49)

$$\varphi(t,\,q) = \ln\sigma\,r_1^{2-q} = \ln\frac{r_1^{2-q}}{(2-q)^2 \int_0^t \alpha\,dt}$$

(3.50)

where r_1 is a reference value of r introduced to make the variables non-dimensional, e.g. the smallest r where concentrations are measured, one obtains:

$$\eta = (2-q)\,\xi + \varphi$$

(3.51)

This formulation is extremely convenient to interpret the observations. Indeed η can easily be calculated as a function of ξ from the survey of the patch of contaminant at any given time. If the solution 3.46 is correct, all points must lie on a straight line. The slope of the line determines q. Figs. 3.2, 3.3 and 3.4 show the diagrams of η plotted from observations of diffusion experiments in the North Sea. The value of q seems to be intermediate between 0 (constant diffusivity) and 1 (Joseph's law), close to 1 during the early stage of the dispersion and approaching zero as the dispersion progresses.

The two-dimensional isotropic solution 3.46 is based on the assumptions of horizontal homogeneity over some region embodying the patch and vertical homogeneity over a thin layer where the contaminant is confined. In that case, the advection velocity is uniform and in axes moving with that velocity the diffusion may be assumed isotropic.

The effect of weak lateral and vertical shears has been examined by Carter

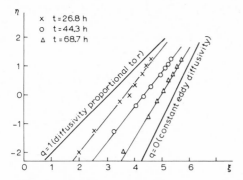

Fig. 3.2. Diagram of η versus ξ from observations during a diffusion experiment in the North Sea (after Joseph et al., 1964). $r_1 = 10$ m.

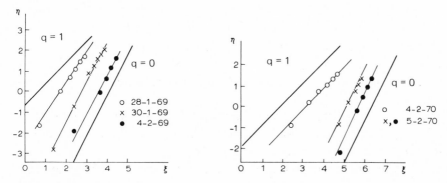

Fig. 3.3. Diagram of η versus ξ from observations during a diffusion experiment in the Southern Bight of the North Sea (after Talbot, 1970). $r_1 = 100$ m.

Fig. 3.4. Diagram of η versus ξ from observations during a diffusion experiment in the North Sea off the Yorkshire coast (after Talbot, 1970). $r_1 = 10$ m.

and Okubo (1965). They consider the mixing from an instantaneous source in an unbounded sea. They take constant but different eddy diffusivities in the directions of the three axes and, allowing for weak lateral and vertical inhomogeneities, they assume a horizontal velocity in the x_1 direction which varies slowly with x_2 and x_3 and may thus be approximated by the first two terms of its Taylor expansion near the origin. They write:

$$u_1 = u(t) + \xi_2 x_2 + \xi_3 x_3 \; ; \quad u_2 = u_3 = 0 \tag{3.52}$$

and

$$\frac{\partial c}{\partial t} + (u + \xi_2 x_2 + \xi_3 x_3) \frac{\partial c}{\partial x_1} = \kappa_1 \frac{\partial^2 c}{\partial x_1^2} + \kappa_2 \frac{\partial^2 c}{\partial x_2^2} + \kappa_3 \frac{\partial^2 c}{\partial x_3^2} \tag{3.53}$$

The solution for an instantaneous point source at $x_1 = x_2 = x_3 = 0$ is found to be:

$$c = \frac{B}{8\pi^{3/2}(\kappa_1\kappa_2\kappa_3)^{1/2}\,t^{3/2}(1+\xi^2t^2)^{1/2}}$$

(3.54)

$$\times \exp-\left[\frac{\{x_1-\int_0^t u\,\mathrm{d}t-\tfrac{1}{2}(\xi_2x_2+\xi_3x_3)t\}^2}{4\kappa_1t(1+\xi^2t^2)}+\frac{x_2^2}{4\kappa_2t}+\frac{x_3^2}{4\kappa_3t}\right]$$

where B is the total amount released and where ξ is a kind of weighted average of the shear given by:

$$\xi^2 = \tfrac{1}{12}\left(\xi_2^2\frac{\kappa_2}{\kappa_1}+\xi_3^2\frac{\kappa_3}{\kappa_1}\right)$$

(3.55)

ξ^{-1} can be interpreted as the time at which the velocity shears begin to affect the mixing significantly.

The mean characteristics of the dispersion predicted by eq. 3.54 are the following:

(1) The surfaces of equal concentration are ellipsoids. The principal axes of the ellipsoids are all in the same direction at any time but their orientation varies with time in the coordinates axes.

(2) The maximum concentration decreases as $t^{-3/2}$ during the initial period $t \ll \xi^{-1}$ and as $t^{-5/2}$ for $t \gg \xi^{-1}$ when the shear effect dominates.

(3) In the period of shear diffusion $t \gg \xi^{-1}$, the patch is greatly elongated in the x_1 direction.

Fig. 3.5 shows the variation with time of the maximum concentration measured by Carter and Okubo (1965) during dye-release experiments in the Cape Kennedy area.

The data distinctly reveals two stages of decay: an initial stage negligibly affected by the inhomogeneity, and a final stage dominated by the shear effect. The corresponding laws of decay are in good agreement with the predictions of the model.

The solutions 3.43 and 3.54 correspond to the very specific situation of an instantaneous point release in a simple current field. Other equally important cases are, for instance, the diffusion from a continuous source, the steady (long-time average) dispersion by residual currents, the distribution of contaminants in stratified waters, etc. Idealized models of such situations and some exact solutions can be found in Crank (1956), Okubo (1971) and Bowden (1972).

Although analytical solutions have proved extremely useful to interpret the observations or to obtain valuable indications to predict the general

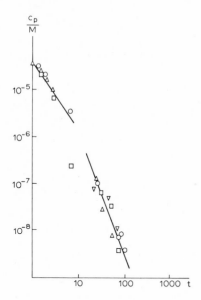

Fig. 3.5. Variation of peak concentration c_p/M (m^{-3}) with time (h) for instantaneous dye-release experiments in the sea off Cape Kennedy (after Carter and Okubo, 1965).

features of the diffusion of contaminants, one must realize that they are limited to simple idealized situations where one can prescribe simple forms to the advection velocity and to the diffusion coefficients.

More detailed descriptions and refined predictions require however the solution of eq. 3.20 taking into account the actual (observed or computed) current pattern and, in most cases, this can only be done numerically.

Numerical techniques often differ from one problem to another, depending in particular on the characteristic time and length scales, the type of sources and the more or less complicated current field.

The numerical methods appropriate to diffusion equations like 3.20 are well documented in the literature and will not be discussed here.

3.5. SHEAR EFFECT IN DEPTH-AVERAGED DISPERSION MODELS

The diffusion equation 3.20 cannot be solved without a prerequisite knowledge of the three components of $\bar{u}$. The analytical solutions discussed in the last section were based on simplified representations of the velocity field. Numerical solutions require the specification of the velocity vectors at a large number of grid points and these must be determined either from the observations or from the results of a preliminary hydrodynamic model.

Most frequently, however, the experimental data are not sufficient to feed

a time-dependent three-dimensional model. In many cases, the only indications available concern the mean or surface value of the horizontal current,
and information is wanting on the vertical component and on the vertical
variations of the velocity vector.

This being so, it is extremely difficult to make any prediction of the
vertical distribution of a contaminant and if it is not, as is assumed in the last
section, confined to a thin layer where the diffusion is essentially horizontal,
the best one can hope for usually is a model for the horizontal dispersion of
the average concentration over the depth.

If the velocity data are derived from a preliminary hydrodynamic model,
the same difficulty arises because time-dependent three-dimensional hydrodynamic models are not operative in the present stage — partly for the same
reason as the deficiency of data affects the quality of their boundary conditions — and the most reliable information is now provided by depth-averaged
hydrodynamic models.

In these circumstances, it seems reasonable firstly to restrict attention to
mean vertical concentrations and develop depth-averaged dispersion models
accorded to the existing hydrodynamic models.

The first step is the vertical integration of the diffusion eq. 3.20. With the
same notations as in Chapter 2 (section 2.2), one defines

(1) the mean vertical concentration:

$$\bar{c} = H^{-1} \int_{-h}^{\zeta} c \, dx_3 \tag{3.56}$$

(2) the deviation from the mean:

$$\hat{c} = c - \bar{c} \tag{3.57}$$

$$\int_{-h}^{\zeta} \hat{c} \, dx_3 = 0 \tag{3.58}$$

As in Chapter 2 (section 2.2), one integrates eq. 3.20 over x_3 from $-h$ to
ζ and one expresses the integrals of the partial derivatives with respect to x_1,
x_2 and t in terms of the derivatives of the integrals by means of eqs. 2.17,
1.44 and 1.45. One obtains (discarding horizontal migration):

$$\frac{\partial (H\bar{c})}{\partial t} + \nabla_h \cdot (H\bar{c}\bar{u}_h) + \nabla_h \cdot \int_{-h}^{\zeta} \hat{c}\hat{u}_h \, dx_3 \tag{3.59}$$

$$= \int_{-h}^{\zeta} \nabla_h \cdot (\kappa \nabla_h c) \, dx_3 + \int_{-h}^{\zeta} S \, dx_3 + \left(\lambda \frac{\partial c}{\partial x_3} - \sigma_3 c\right)_{x_3 = \zeta} -- \left(\lambda \frac{\partial c}{\partial x_3} - \sigma_3 c\right)_{x_3 = -h}$$

One verifies here again the general remark made in Chapter 1 that integration over space incorporates some of the boundary conditions in the evolution equations. The last two terms in the right-hand side of eq. 3.59 represent the fluxes at the air—sea interface and at the bottom. Classical semi-empirical formulas for these fluxes have been discussed in Chapter 1. It is assumed here that they are known (as functions of atmospheric conditions, etc.).

The last three terms in the right-hand side of eq. 3.60 combine and constitute the total input in a water column of unit base. In brief, one shall write:

$$H\Lambda = \int_{-h}^{\zeta} S \, dx_3 + \left(\lambda \frac{\partial c}{\partial x_3} - \sigma_3 c\right)_{x_3 = \zeta} - \left(\lambda \frac{\partial c}{\partial x_3} - \sigma_3 c\right)_{x_3 = -h} \tag{3.60}$$

introducing the mean input Λ, a given function of x_1, x_2 and t.

Eliminating the time derivative of H by eq. 2.22 and dividing by H, one can write eq. 3.59 in the simple form:

$$\frac{\partial \bar{c}}{\partial t} + \bar{u}_h \cdot \nabla_h \bar{c} = \Lambda + \Sigma + T \tag{3.61}$$

where:

$$T = \int_0^1 \nabla_h \cdot (\kappa \nabla_h c) \, d\eta \tag{3.62}$$

$$\Sigma = -H^{-1} \nabla_h \cdot (H \int_0^1 \hat{u}_h \hat{c} \, d\eta) \tag{3.63}$$

and η is a new variable defined by

$$\eta = H^{-1}(x_3 + h) \tag{3.64}$$

Again the non-linear terms are found to give two contributions, the first of which is the product of the means while the second, Σ, is related to the mean product of the deviations.

One can see that Σ is a result of the vertical shear. It would be zero if the velocity was uniform over the depth. (This however cannot be the case because, as explained in Chapter 2, the velocity is maximum at the surface and vanishes at the bottom.)

One can argue, as in Chapter 2, that Σ will contribute to enhance the dispersion. This effect is called the *shear effect.*

A note of precaution is required here. In section 3.4, in a slightly different situation, the name "shear effect" has been introduced in connection with the action of weak current inhomogeneities on the horizontal dispersion of a contaminant confined to a thin layer of fluid. In Chapter 2, the name appeared in association with the derivation of depth-averaged equations and it is applied in the same capacity here.

In fact, the appellation "shear effect" is widely used to denote similar but not identical phenomena. In this context, it is meant in the following sense: space-average concentrations (e.g., over the depth or over the cross-section) are governed by equations which are derived from the three-dimensional ones by space integration. In this process, the quadratic convection terms give two contributions, the first of which represents the advection by the mean motion while the second contains the mean product of deviations around the means and contributes to the dispersion.

This effect has been described by several authors in pipes, channels and estuaries where, after integration over the cross-section, the flow — steady or oscillating — is essentially in one direction (Taylor, 1953, 1954; Elder, 1959; Bowden, 1965).

In shallow waters, it is generally sufficient to consider the mean concentrations over the depth but, out at sea, no further averaging is possible and the dispersion mechanism is fundamentally two-dimensional. The models valid for estuarine diffusion have been generalized by Nihoul (1972, 1973b) to account for transient wind-changing currents and rotating tidal currents.

In Chapter 2, the combined effect of the shear and turbulent dispersions was simply taken into account by adjusting the diffusion term and modifying the values of the dispersion coefficients.

Although such an approximation may be justified for hydrodynamic models concerned with the circulation over extended regions of the sea, it must be refined for dispersion models which very often deal with the diffusion of a contaminant over a limited region around a point source. While in the first case the coarseness of the grid and the flexibility of the numerical viscosity concept can easily incorporate the shear effect, in the diffusion case the same approach would be too crude for many problems and it is indispensable to have better estimates of Σ.

To evaluate the shear effect in passive dispersion models, one proceeds as follows (Nihoul, 1972, 1973b). Substracting eq. 3.61 from 3.20, one gets:

$$\frac{\partial \hat{c}}{\partial t} + \bar{u}_h \cdot \nabla \hat{c} + \hat{u}_h \cdot \nabla \hat{c} + u_3 \frac{\partial \hat{c}}{\partial x_3} + \Sigma - \nabla_h \cdot (\kappa \nabla_h c) + \mathrm{T} + \hat{u}_h \cdot \nabla \bar{c}$$

$$= \frac{\partial}{\partial x_3} \left(\lambda \frac{\partial \hat{c}}{\partial x_3} - \sigma_3 \hat{c} \right) + S - \Lambda \tag{3.65}$$

It is reasonable to assume that the deviation $\hat{c}$ is much smaller than the mean concentration $\bar{c}$ while the velocity deviation $\hat{u}_h$ can be a substantial fraction of $\bar{u}_h$ over a large part of the water column as a result of the velocity profile imposed by the boundary conditions.

It is generally accepted that the vertical advection and the turbulent diffusion residue can be neglected as compared to the horizontal advection. For instance, one estimates (introducing the diffusion velocity P):

$$T - \nabla_h \cdot (\kappa \nabla_h c) \sim 0 \left(\frac{P\hat{c}}{\ell} \right) \ll \hat{u}_h \cdot \nabla \bar{c} \sim 0 \left(\frac{\bar{u}_h \bar{c}}{\ell} \right) \tag{3.66}$$

Under these conditions, all terms in the left-hand side of eq. 3.65 are small compared to the last one, and one may write:

$$\hat{u}_h \cdot \nabla \bar{c} \sim \frac{\partial}{\partial x_3} \left(\lambda \frac{\partial \hat{c}}{\partial x_3} - \sigma_3 \hat{c} \right) + S - \Lambda \tag{3.67}$$

The physical meaning of this equation is clear; weak vertical inhomogeneities are constantly created by the inhomogeneous convective transfer of the admixture and they adapt to this transfer in the sense that the effects of convection, transverse diffusion and unequal migration are balanced for them.

Eq. 3.67 can be used to calculate $\hat{c}$ in terms of the gradient of the mean concentration $\bar{c}$. Multiplying the result by $\hat{u}_h$ and integrating over depth, one obtains thus an estimate of the shear effect.

The result turns out to be fairly simple if one assumes, following Bowden (1965), that, at least in shallow seas, one may write:

$$\hat{u}_h = \bar{u}_h \varphi(\eta) \tag{3.68}$$

$$\lambda = \bar{u}_h H g(\eta) \tag{3.69}$$

Eq. 3.67 becomes:

$$(\bar{u}_h \cdot \nabla \bar{c})\varphi = -\frac{\sigma_3}{H} \frac{\partial \hat{c}}{\partial \eta} + \frac{\bar{u}_h}{H} \frac{\partial}{\partial \eta} \left(g \frac{\partial \hat{c}}{\partial \eta} \right) + S - \Lambda \tag{3.70}$$

Integrating twice, multiplying by $\hat{u}_h$ and averaging over depth, one can calculate Σ. In the case of sedimentation ($\sigma_3 < 0$), assuming zero flux at the free surface and neglecting small-order terms, one finds*:

$$\Sigma = H^{-1} \nabla \cdot H \left[\gamma_1 \frac{H}{\bar{u}_h} \bar{u}_h (\bar{u}_h \cdot \nabla \bar{c}) - \gamma_2 \frac{\sigma_3}{\bar{u}_h} \bar{c} \bar{u}_h \right] \tag{3.71}$$

where

$$\bar{u}_h = \|\bar{u}_h\| \tag{3.72}$$

---

* The case $\sigma_3 > 0$ can be treated in the same way with only trivial modifications of the algebra.

$$\gamma_1 = -\int_0^1 \varphi \, d\eta \int_0^\eta g^{-1} \, d\xi \int_1^\xi \varphi \, d\beta \tag{3.73}$$

$$\gamma_2 = \int_0^1 \varphi \, d\eta \int_0^\eta g^{-1} \, d\xi \tag{3.74}$$

Among the small terms neglected in eq. 3.71 is the contribution from the successive integration of the last two terms in the right-hand side of eq. 3.70. This contribution is indeed of the order $(H/L)\Lambda$ where L is a characteristic length of horizontal variations and may be expected to be much larger than the depth. Hence, after substitution in eq. 3.61, this term can be neglected as compared to Λ.

The coefficients γ_1 and γ_2 are of order 1. Their particular values depend upon the form of the functions φ and g. They can be estimated for instance by using Van Veen's power law ($u = 1.2 \, \bar{u}\eta^{0.2}$) inferred from a large number of experiments in Dutch estuaries and in the Straits of Dover. Sometimes, a better fit is obtained with a combined log-parabolic profile (Bowden and Fairbairn, 1952). Calculating γ_1, assuming a log-parabolic profile and a ratio of eddy diffusivity to eddy viscosity equal to 1.4 (Ellison, 1957), Nihoul found a value of 0.45 which compared very well with the observations (Nihoul, 1972).

The last term in the right-hand side of eq. 3.71 represents a contribution to the shear effect due to the migration. This contribution is negligible in most cases because the migration velocity σ_3 is usually rather small. It may become more important though in sedimentation problems when flocs are formed or when large-size solid lumps are released as observed for instance in some dumping grounds off the Belgian coast. In this case, its effect is more a reduction of the advection than an increase of the dispersion. This is easy to understand from a physical point of view. If the sedimentation velocity is important, the particles — although a high friction velocity may prevent their deposition and maintain them in suspension — will tend to concentrate in a region closer to the bottom where the velocity is smaller than the mean velocity. The second term in the left-hand side of eq. 3.61 thus overestimates the real advection of those particles and this is corrected by the first part of the shear effect. (Note that σ_3 is negative in the case of sedimentation.) Of course the argument does not apply if the concentration becomes too large as the whole model is based on the hypothesis of weak inhomogeneities.

The first term in the right-hand side of eq. 3.71 may be interpreted as a dispersion term and appears as a natural generalization to variable depth and two-dimensional horizontal flows of Bowden's results (Bowden, 1965).

Combining eqs. 3.61 and 3.71 and introducing a new horizontal eddy diffusivity $\tilde{\kappa}$ such that:

$$T \sim \nabla \cdot \tilde{\kappa} \nabla \bar{c} \tag{3.75}$$

one obtains:

$$\frac{\partial \bar{c}}{\partial t} + \bar{u}_h \cdot \nabla \bar{c} + H^{-1} \nabla \cdot \left(\gamma_2 \frac{H\sigma_3}{\bar{u}_h} \bar{c} \, \bar{u}_h \right)$$

$$= \Lambda + H^{-1} \nabla \cdot \left[\gamma_1 \frac{H^2}{\bar{u}_h} \bar{u}_h (\bar{u}_h \cdot \nabla \bar{c}) \right] + \nabla \cdot \tilde{\kappa} \nabla \bar{c} \tag{3.76}$$

This equation has been used extensively in modelling the dispersion of pollutants in the Southern Bight of the North Sea. The velocity $\bar{u}_h$ and depth H are either measured or calculated by a separate numerical model combining the effects of winds and tides. The coefficients γ_1, γ_2 and $\tilde{\kappa}$ can be estimated theoretically but they are as easily determined by numerical simulation if, as often, observations are available for reference. The calculated values agree well with the theoretical predictions.

Examining the second term in the right-hand side of eq. 3.76 one can see that the shear effect produces a diffusion in the direction of the instantaneous velocity with an apparent diffusivity of the order of $\bar{u}_h H$. In regions of strong tidal currents, the isotropic turbulent dispersion is usually much smaller as the eddy diffusivity $\tilde{\kappa}$ can be two orders of magnitude smaller than uH. After one or two tidal periods, there results an enhanced dispersion in the direction of the maximum tidal velocity. Rapid variations of depth and strong winds can of course modify the situation but the tendency remains and the patches of pollutants have very often an elongated shape with a maximum dispersion roughly in the direction of the maximum current (Fig. 3.6).

This prediction of the model is well verified by the observations (Nihoul, 1971).

One of the consequences of the cogent influence of the shear effect is that the second-order operator in eq. 3.76 is nearly parabolic; the dominant terms being:

$$\frac{H}{\bar{u}_h} \left(\bar{u}_1^2 \frac{\partial^2 \bar{c}}{\partial x_1^2} + \bar{u}_2^2 \frac{\partial^2 \bar{c}}{\partial x_2^2} + 2 \bar{u}_1 \bar{u}_2 \frac{\partial^2 \bar{c}}{\partial x_1 \, \partial x_2} \right) \tag{3.77}$$

As one might expect, this created numerical difficulties and required a careful test of several techniques before a reliable one was perfected (Adam and Runfola, 1972).

In many cases, when one is only interested in a prediction of the pollutant's concentration every one or two tidal periods, it is advantageous to

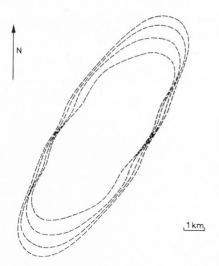

Fig. 3.6. Simulation of a dye-release experiment in the North Sea (after Adam and Run-fola, 1972). Position: 51° 20′N 1° 34′E. Curves: 1/50 of initial central concentration — 48 h — 72 h — 96 h — 108 h after release.

average first eq. 3.76 over a tidal period. This generally reduces considerably the numerical difficulties, and it is easy to see why it should be so. Indeed, if in eq. 3.77 $\bar{c}$ was independent of time, the integration over a tidal period would make the operator elliptic by virtue of Schwartz's inequality. Of course $\bar{c}$ is not independent of time in general and in particular in the case of dispersion from an instantaneous source simulating a dumping.

However, it is usually possible to separate $\bar{u}_h$ and $\bar{c}$ into slowly varying parts (over a tidal period) and rapidly oscillating parts (with the tidal period) and to apply the averaging technique of Krylov Bogoliubov and Mitropolsky with the same benefit (Nihoul, 1972).

It is illuminating to consider the following simple case:

(1) H is a constant (e.g., the surface elevation is negligible as compared to the depth and the latter is constant over the region of interest).

(2):

$$\bar{u}_h = ae_1 \cos \omega t + be_2 \sin \omega t \tag{3.78}$$

where ae_1 and be_2 are two constant vectors. Eq. 3.78 postulates that, over the region of interest, the velocity vector rotates with tides and the tidal ellipse is the same at all points.

(3) Eddy diffusion may be neglected as compared to shear effect.

(4):

$$\bar{c} = \bar{c}_m + \bar{c}_p \tag{3.79}$$

where $\bar{c}_m$ varies slowly over a tidal period while $\bar{c}_p$ oscillates with tides and indeed is directly related to $\bar{u}_h$ (Nihoul, 1972).

(5) Vertical migration is negligible.

(6) $\Lambda = 0$ in the case of an instantaneous point release. Then, integrating eq. 3.76 over a tidal period, one obtains:

$$\frac{\bar{c}_m(t+T)-\bar{c}_m(t)}{T} = T^{-1}\int_t^{t+T}\left(-\bar{u}_\alpha\frac{\partial\bar{c}}{\partial x_\alpha} + \gamma_1\frac{H}{\bar{u}_h}u_\alpha u_\beta\frac{\partial^2\bar{c}}{\partial x_\alpha\,\partial x_\beta}\right)dt \quad (3.80)$$

Since $\bar{c}_m$ is assumed to vary slowly in a tidal period, the finite difference in the left-hand side may, in first approximation, be replaced by the derivative, and the integral in the right-hand side may be evaluated keeping $\bar{c}_m$ constant. The terms containing $\bar{c}_p$ are then zero by symmetry.

One obtains, thus:

$$\frac{\partial c_m}{\partial t} \sim \nu_1\frac{\partial^2 c_m}{\partial x_1^2} + \nu_2\frac{\partial^2 c_m}{\partial x_2^2} \quad (3.81)$$

where:

$$\nu_1 = \frac{2\alpha\,Ha}{\pi}\,B(k) \quad (3.82)$$

$$\nu_2 = \frac{2\alpha\,Ha}{\pi}\,\frac{b^2}{a^2}\,D(k) \quad (3.83)$$

where B and D denote the complete elliptic integrals:

$$B(k) = \int_0^{\pi/2}\cos^2\theta\,(1-k^2\sin^2\theta)^{-1/2}\,d\theta \quad (3.84)$$

$$D(k) = \int_0^{\pi/2}\sin^2\theta\,(1-k^2\sin^2\theta)^{-1/2}\,d\theta \quad (3.85)$$

and where k is the eccentricity of the tidal ellipse:

$$k^2 = 1-\frac{b^2}{a^2} \quad (3.86)$$

The solution of eq. 3.81 is:

$$\bar{c}_m = \frac{A}{\sqrt{\nu_1\nu_2}\,t}\,\exp\left(-\frac{x_1^2}{4\nu_1 t} - \frac{x_2^2}{4\nu_2 t}\right) \quad (3.87)$$

where the constant A depends on the quantity released at $t = 0$.

Hence the curves of equal concentration at time t are ellipses with equations:

$$\frac{x_1^2}{4\nu_1 t} + \frac{x_2^2}{4\nu_2 t} = \ln \frac{\bar{c}_0}{\bar{c}_m} \tag{3.88}$$

where $\bar{c}_0$ denotes the concentration at the center of the patch. The isoconcentration ellipses and the tidal ellipses have parallel axes. Their eccentricities ϵ and k are related by:

$$\epsilon^2 = 1 - (1 - k^2) \frac{D(k)}{B(k)} \tag{3.89}$$

It is instructive to compare these results with those of an experiment made by Talbot (1970) in January 1969 with the dye tracer rhodamine B in the southern North Sea. Sixty-eight hours after release, Talbot reported a patch of rhodamine of roughly elliptical shape. Fitting an ellipse to the tidal vector diagram at the center of the patch and computing the curves of equal concentrations by eqs. 3.88 and 3.89, one finds that the isoconcentration ellipses predicted by the model agree very well in *orientation*, *shape* and *size* (for $\gamma_1 \sim 0.45$) with Talbot's observations (Nihoul, 1972; Fig. 3.7).

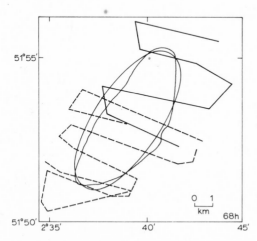

Fig. 3.7. Comparison between observed and predicted shape of a patch of rhodamine B, 68 h after release. The experimental curve is the irregular curve drawn by Talbot (1970) (the broken piecewise straight lines are the ship's trajectories). The theoretical curve is the regular ellipse predicted by the simplified model (Nihoul, 1972).

REFERENCES

Adam, Y. and Runfola, Y., 1972. *Programme national sur l'environnement physique et biologique, projet mer, N14.* Department of Scientific Policy, Brussels.

Bowden, K.F., 1965. *J. Fluid. Mech.*, 21: 83.

Bowden, K.F., 1972. *Mém. Soc. R. Sci. Liège*, 2: 67.

Bowden, K.F. and Fairbairn, L.A., 1952. *Proc. R. Soc. Lond.*, 214A: 371.

Carter, H.H. and Okubo, A., 1965. *Chesapeake Bay Inst., Tech. Rep.*, Ref. 65—2.

Cooper, J.W. and Stommel, H., 1968. *J. Geophys. Res.*, 73: 5849.

Crank, J., 1956. *The Mathematics of Diffusion.* Oxford University Press, Oxford, 356 pp.

Elder, J.W., 1959. *J. Fluid. Mech.*, 5: 544.

Ellison, T.M., 1957. *J. Fluid. Mech.*, 2: 456.

Grant, H.L., Moilliet, A. and Vogel, W.M., 1968. *J. Fluid. Mech.*, 34: 443.

Joseph, J. and Sendner, H., 1958. *Dtsch. Hydrogr. Z.*, 11: 49.

Joseph, J. and Sendner, H., 1962. *J. Geophys. Res.*, 67: 3201.

Joseph, J., Sendner, H. and Weidemann, H., 1964. *Dtsch. Hydrogr. Z.*, 17: 57.

Kent, R.E. and Pritchard, D.W., 1959. *J. Mar Res.*, 18: 62.

Kullenberg, G., 1970. *K. Vetensk. Vitterhets Samhället, Göteborg, Ser. Geophys.*, 2.

Mamayev, O.I., 1958. *Izv. Akad. Nauk S.S.S.R., Geofiz.*, 7: 870.

Munk, W.H., 1966. *Deep-Sea Res.*, 13: 707.

Munk, W.H. and Anderson, E.R., 1948. *J. Mar. Res.*, 7: 276.

Nihoul, J.C.J., 1971. *Proc. North Sea Sci. Conf., Aviemore*, 1: 89.

Nihoul, J.C.J., 1972. *Bull. Soc. R. Sci. Liège*, 10: 521.

Nihoul, J.C.J., 1973a. *Mém. Soc. R. Sci., Liège*, 4: 115.

Nihoul, J.C.J., 1973b. *Proc. IUTAM—IUGG Symp. Turbulent Diffusion in Environmental Pollution, 2nd, Charlottesville, 1973.*

Okubo, A., 1968. *Chesapeake Bay Inst., Tech. Rep.*, 38, Ref. 62—22.

Okubo, A., 1971. In: D.W. Wood (Editor), *Impingement of Man on the Oceans.* Interscience, New York, N.Y., pp. 89—168.

Ozmidov, R.V., 1965. *Izv. Atmos. Oceanic Phys. Ser.*, 1: 257.

Stewart, R.W., 1959. *Adv. Geophys.*, 6: 303.

Stewart, R.W. and Grant, H.L., 1962. *J. Geophys. Res.*, 67: 3177.

Stommel, H. and Fedorov, K.N., 1967. *Tellus*, 19: 306.

Takenouti, Y., Nanniti, T. and Yasui, M., 1962. *Oceanogr. Mag.*, 13: 89.

Talbot, J.W., 1970. *Proc. FAO Tech. Conf. Marine Pollution, Rome, 1970.*

Taylor, G.I., 1953. *Proc. R. Soc. Lond.*, 219A: 186.

Taylor, G.I., 1954. *Proc. R. Soc. Lond.*, 223A: 446.

Van Veen, J., 1938. *J. P.-V. Cons. Int. Explor. Mer*, 13: 7.

Woods, J.D., 1968a. *Meteorol. Mag.*, 97: 65.

Woods, J.D., 1968b. *J. Fluid. Mech.*, 32: 791.

CHAPTER 4

INTERACTION MODELS

Jacques C.J. Nihoul

4.1. INTRODUCTION

Dealing with chemical or biological variables, it often happens that insufficient information is available to determine the form of the interaction terms as a function of space and time and that one must resort to statistical observations over some (large) region of space. The whole region is then regarded as a homogeneous system or "box", the main organs of which are interactions. This is often the case in fishery dynamics where the relative rate of change in biomass of the fished stock is calculated from the average rate of recruitment, growth and natural mortality (interaction effects), the fishing effort (external sink effect) and the migration in and out of the region due to variations in external environmental factors (a boundary source or sink effect).

In a more refined approach to the dynamics of the evolution and interaction of populations, one may separate the region into several niches where similar environmental conditions prevail and associate a homogeneous system to each niche.

The early mathematical models of plankton production for instance are all box models based on the assumption of horizontal homogeneity. They are in most cases further simplified by assuming either a steady state (when the author is interested in the depth variations) or a vertical homogeneity (when the author is interested in the annual cycle).

In Steele's model of productivity in the northern North Sea (Steele, 1958), complete homogeneity is assumed in each of two horizontal layers. Vertical exchange between these layers is accounted for by introducing a control parameter, the mixing coefficient m.

Box models are in fact answerable to the same philosophy as depth-averaged hydrodynamic models.

Marine systems are vast and complex. Modelling, on the other hand, is limited by the shortage of experimental data — data are required to determine the parameters, the boundary and initial conditions, etc. — and by the restricted capacity of the available computers.

It is thus often necessary to simplify the system one wishes to model and, as explained in Chapter 1, one naturally tries to reduce the size of the system and, in particular, the size of the support by integrating over one or several space coordinates.

When one is particularly interested in the time evolution of biological populations or chemical concentrations, it seems reasonable in a first approach, to perform a complete space integration over the region of interest.

In a more refined way, the region can be divided into several individual boxes characterized by their mean or integral properties. A model is then constructed for each box taking into account, in the inputs and outputs, the flows of material from one box to another.

The dynamics of the box-averaged variables is then governed by the interactions between them and, for that reason, box models are generally called *interaction models.*

4.2. BOX MODEL EQUATIONS

Eq. 1.19 can be written, using the eddy diffusivity concept:

$$\frac{\partial r_\alpha}{\partial t} = Q_\alpha + I_\alpha + \nabla \cdot (\kappa \nabla r_\alpha - r_\alpha \boldsymbol{\sigma} - r_\alpha \boldsymbol{u}) \tag{4.1}$$

If $\mathcal{V}$ is a fixed volume, the integral over $\mathcal{V}$ of the last term in the right-hand side can be transformed by Gauss's theorem into an integral over the surface encompassing $\mathcal{V}$*. This integral is thus determined by the boundary conditions on the surface.

Hence in a box model, the contributions from advection, migration and diffusion reduce to the flux through the boundary of the volume $\mathcal{V}$ of the box. On the sections of the boundary which coincide with actual frontiers of the total system (like the air—sea interface or the bottom), the flux is determined by the general boundary conditions imposed on the system; on sections of the boundary which separate the box from other adjacent boxes, the **flux depends on what happens in the boxes around it.** Thus if the system's support is divided into several boxes, the model for each box will be coupled with the models for the others through the mutual exchanges at their common boundaries. One must then solve the equations for all the boxes simultaneously.

---

* If different eddy diffusivities are used in different directions, the conclusion is the same. The diffusion term is no longer a divergence but is the sum of partial derivatives with respect to x_1, x_2 and x_3, and the integrals of these depend only on the conditions at the boundary.

As pointed out in Chapter 1, it may be tempting to multiply the boxes to ensure a better coverage of the system but, in so doing, one loses the simplicity which space integration has allowed. Indeed, while effectively reducing the support (to the sole time variable), one increases at the same time the scope of the system and if there are initially n state variables and we consider r boxes, we come out with what is in fact a gigantic model of rn unknowns.

Hence box models are particularly useful when we can assume reasonable spatial homogeneity and obtain a satisfactory description of the system with a limited number of boxes.

In the limit, if the spatial variations are not essential, one will treat the whole system as a single box. The fluxes in or out of the box are then completely determined by the boundary conditions.

In the following, to emphasize the interaction processes, we will restrict attention to the case of a single box model. Box averages will be noted by a bar and the total mean input will be written S_α, for simplicity, i.e.:

$$\mathcal{V}^{-1} \int_{\mathcal{V}} [Q_\alpha + \nabla \cdot (\kappa \nabla r_\alpha - r_\alpha \, \boldsymbol{\sigma} - r_\alpha \boldsymbol{u})] \; \mathrm{d}\mathcal{V}$$

$$= \mathcal{V}^{-1} \int_{\mathcal{V}} Q_\alpha \, \mathrm{d}\mathcal{V} + \mathcal{V}^{-1} \int_{\Sigma} (\kappa \nabla r_\alpha - r_\alpha \, \boldsymbol{\sigma} - r_\alpha \boldsymbol{u}) \cdot \mathrm{d}\boldsymbol{\Sigma} \qquad (4.2)$$

$$= S_\alpha$$

Averaging eq. 4.1, one then obtains the differential equation:

$$\frac{\mathrm{d}\bar{r}_\alpha}{\mathrm{d}t} = S_\alpha + \bar{I}_\alpha \qquad (4.3)$$

In eq. 4.3, S_α is given and represents the rate of production (or destruction) of $\bar{r}_\alpha$ by external effects (dumping, fishing, etc.) or by flows and fluxes through the boundaries.

$\bar{I}_\alpha$ represents the rate of production (or destruction) of r_α by chemical, biochemical or ecological interactions. $\bar{I}_\alpha$ must be given appropriate form in terms of the state variables and control parameters.

To this effect one must first identify the interactions which play a significant role in the marine system. One is guided here by the results of laboratory studies and theoretical analyses and by the inspection of the series of data provided by the observations. One must then identify the nature of the interactions and formulate the laws of interactions in a form appropriate to the model. Valuable information is again obtained from theory and laboratory experiments. It may be necessary in addition to run separate simulation models to test tentative formulas, determine reaction rates and other coefficients and check the applicability to the complex marine system of the simple rules which have emerged from the observations in the laboratory conditions.

The identification of the significant interactions reveals the variables which are interrelated and contribute to the different reaction rates $\bar{I}_\alpha$. Some of these variables which obviously pertain to the system's environment can immediately be taken as control parameters and given empirical forms. The final selection of the state variables however cannot be made before the interaction laws are investigated and one knows to what extent reliable evolution equations can be formulated. In some cases, the role of certain important variables cannot be expressed in a convenient mathematical form and we must consider them as control parameters necessitating a further input of empirical data into the model.

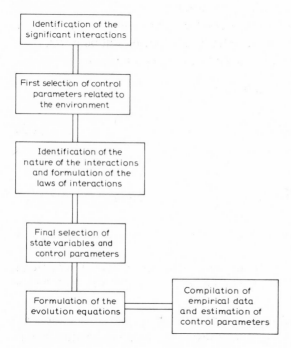

Fig. 4.1. Successive steps in the construction of an interaction model.

The different steps in the construction of an interaction model are sketched in Fig. 4.1.

4.3. THE INTERACTION RATES

Although theoretical research, laboratory experiments and field observations have allowed the chemists and the biologists to get a valuable insight into the laws of interactions, one must admit that they have not been so far sufficient — nor, indeed, entirely appropriate — to determine the mathematical form of the interaction rates $\bar{I}_\alpha$ in accordance with the models desiderata.

There are for that several reasons:

(1) The oceanographic surveys and the interpretation of the observations are rarely, and have almost never been in the past, conceived to provide information for modelling.

One often discovers that in one particular survey, one or several essential parameters for modelling have not been measured and that the data are accordingly rather inconclusive for the purpose. In general, the interpretation of the observations is limited to a simple ordering of the results for presentation to the scientific community, and local or occasional effects often mask the universal rule one should be looking for. Transposing data from one area to another is then always hazardous.

(2) Laboratory experiments and theoretical research are limited by the very conditions of their design. In general they are planned to investigate one particular aspect of a phenomenon. To emphasize the process under examination, one usually pays a special attention to extreme situations where one can guarantee that this process dominates, the experimental set-up being such that rival processes are excluded by artificially maintaining competing variables at a known constant level. Hence theoretical and experimental studies, although they serve to elucidate the role of some important factors, are not directly applicable to real, complex, in situ interactions.

(3) As discussed in Chapter 1, faced with the necessity of simplifying the model system, one naturally tries to abandon detailed descriptions and, in a first approach, to restrict attention to the aggregate properties of the most essential compartments in state space. In this respect, the state variable r_α may refer to the total mass of nutrients per unit volume or to the specific biomass of phytoplankton while laboratory studies concentrate on one particular compound of phosphorus or one particular alga. In Chapter 1, the word "translocation" was introduced to denote a transfer from one compartment to another. It is obvious that the dynamics of translocations is the result of an intricate web of chemical or ecological reactions and is often governed by laws which are not simple consequences of chemical kinetics or ecology.

An interesting example was given by Kelly (1971) who pointed out that the overall behaviour of lumped species components of the system may be quite different from the response of individual species of organisms. Fig. 4.2, for example, reproduced from Kelly (1974), shows the rate of nutrient uptake by algae as a function of the temperature of the system.

Although the rate of each species is found to have a definite maximum, the total rate increases exponentially.

Interaction models are thus guided by field observations and laboratory

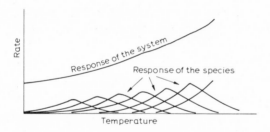

Fig. 4.2. Rate of nutrient uptake by algae as a function of temperature (after Kelly, 1974).

experiments but the final specification of the interaction terms $\bar{I}_\alpha$ requires an oriented exploitation of the data.

Cross-correlation studies of series of observations contribute to identify the cogent interactions. Test formulas — inspired by laboratory and theoretical research but not construed by them — are fitted to the data by least-squares or other methods. Specially designed simulation programs are run to check the realism of simple mathematical forms and ascertain the values of coefficients.

In most cases, the conclusions enforce unvarnished expressions which are most easily adapted to the description of the experimental data and the conduct of the model.

(1) The simplest possible relationship is the linear one, i.e.:

$$\bar{I}_\alpha = k_{\alpha\beta}\bar{r}_\beta \tag{4.4}$$

where a sum is made, if more than one constituent is involved, on the repeated subscript β and where the $k_{\alpha\beta}$'s are eventually (known) functions of time.

Although this may seem a rather crude approximation, it has been used extensively in ecology — where the biologist commonly speaks of the exponential growth or decay of a species[*] — and it can be justified if one restricts attention to the evolution of the system in the neighbourhood of some (known) reference state. Eq. 4.4 can then be regarded as representing the first terms of the Taylor expansion of $\bar{I}_\alpha$ around the state of reference.

In other situations, eq. 4.4 may hold because the reaction rate may be assumed proportional to the state variables even if it is a very complicated

---

[*] The exponential growth or decay, in its simple form, is the solution of an evolution equation of the type

$$\frac{d\bar{r}_\alpha}{dt} = k\bar{r}_\alpha$$

which is a special case of eqs. 4.3 and 4.4.

function of other variables which, in the particular problem under consideration, are taken as control parameters and given empirical forms.

For instance, in Steele's model of primary production (Steele, 1958, 1962) later modified by Lassen and Nielsen (1972) and adapted to hybrid computation by Droissart et al. (1973), the rate of destruction of phytoplankton is assumed proportional to the phytoplankton's biomass. It is really a complicated function of several variables like turbulent mixing, zooplankton grazing, etc., but in the model these are control parameters determined empirically and their effects are combined in one single function of time which appears in the evolution equation as the coefficient of proportionality in the destruction rate.

(2) The simplest approximation which comes next in the hierarchy of difficulty is the bilinear-quadratic law of the form:

$$\bar{I}_\alpha = k_{\alpha\beta\gamma} \bar{r}_\beta \bar{r}_\gamma + k_{\alpha\beta} \bar{r}_\beta \tag{4.5}$$

where a sum is made, if required, on repeated subscripts.

In some cases, expressions like 4.5 appear naturally as first- and second-order contributions from a Taylor expansion around some known reference state.

In any case, the logic of the non-linear terms is clear. The interactions between prey and predator or between two rival species competing for the same food depend, in the first instance, on the number of binary encounters effective between them, and this is proportional to the product of the two populations (Lotka, 1956).

Population growth is often described by the logistic equation (Slobodkin, 1966; Odum, 1972):

$$\frac{d\bar{r}_\alpha}{dt} = a\bar{r}_\alpha \left(1 - \frac{\bar{r}_\alpha}{r_0}\right) \tag{4.6}$$

where r_0 is the carrying capacity of the environment (the maximum mass or number of individuals that can be maintained by the environment under equilibrium conditions).

Eq. 4.6 is a particular case of 4.5 if a is a known function of time.

Different forms of the logistic equation have been proposed where the coefficients of the linear and quadratic terms depend upon the availability of food, the increased respiration caused by the use of energy resources in detoxification or other functions, etc.

Still if these effects are measured by control parameters, the reaction rate $\bar{I}_\alpha$ remains quadratic in the state variable $\bar{r}_\alpha$.

(3) One expects that the rate of translocation of a given constituent depends on its concentration in the compartment from which it is taken. The more there is available, it would seem, the faster the transfer will progress.

Although this may be the case initially, the transfer rate cannot grow forever. There is in general a physical limitation on the amount of material that can be taken up per unit time. It is the case, in particular, for the uptake of nutrients by the phytoplankton, the feeding of zooplankton, bacteria, fish, etc. (e.g. Parsons and Le Brasseur, 1970; Fuhs et al., 1972).

To describe this effect, Odum (1972) has suggested a Michaelis-Menten-Monod law of the form:

$$\bar{I} = a \frac{\bar{r}}{b + \bar{r}} \tag{4.7}$$

where $\bar{I}$ is the rate of translocation, $\bar{r}$ is the concentration of the constituent being transferred, and where a and b depend on the receiving compartment and may be functions of other state variables or control parameters expressing for instance the influence on the transfer rate of environmental factors (temperature, toxic materials, etc.).

It is illuminating to consider two limiting cases of eq. 4.7.

If $\bar{r} \ll b$, one can write:

$$\bar{I} = \frac{a}{b} \frac{\bar{r}}{1 + \frac{\bar{r}}{b}} \sim \frac{a}{b} \bar{r} \left(1 - \frac{\bar{r}}{b}\right) \tag{4.8}$$

In this case, the Michaelis-Menten-Monod law can be approximated by a quadratic law (or even a linear law if one is prepared to neglect $\bar{r}/b$ as compared to 1 in the parentheses).

If, on the other hand, $\bar{r} \gg b$, one can write:

$$\bar{I} = a \frac{1}{1 + \frac{b}{\bar{r}}} \sim a \left(1 - \frac{b}{\bar{r}}\right) \tag{4.9}$$

In the limit, when $b/\bar{r}$ is very small, the interaction rate reduces to a and becomes independent of the concentration $\bar{r}$.

Pichot (1973) gives two examples of this situation in the Ostend Bassin de Chasse: (1) the uptake of nutrients by the phytoplankton depends on nitrites and nitrates but not on the phosphate concentration which is more than 20 times the saturation value (according to Sen Gupta, 1969, this could be a general feature of coastal marine ecosystems); and (2) the grazing effect is not affected by the biomass of the phytoplankton which exceeds the value corresponding to the maximum feeding rate of the zooplankton.

4.4. PRODUCTIVITY MODELS

One of the most important aspects of the marine system's dynamics is the chain of interactions by which the phytoplankton consumes the nutrients, the zooplankton feeds on the phytoplankton, the carnivores on the zooplankton, and through a complicated web of prey—predator relationships, life is maintained in the sea from the microscopical algae to the big fish which constitute an important resource of food for man.

The first attempts to model the food chain have naturally concentrated on the first trophic levels and have tried to simulate the evolution of plankton populations under the influence of observed or predicted nutrient concentrations.

As more data became available and the numerical techniques were better understood, the initial studies developed into more sophisticated models including additional state variables like bacterial numbers or carnivores and incorporating better the influence of environmental factors like temperature, light extinction, etc. Several recent examples are discussed by Dugdale and Steele (Chapters 9 and 10).

Attention will be restricted here to two examples which will serve to illustrate the techniques, the virtues and the limitations of the interaction models.

In a series of publications, Steele (1956, 1958, 1959, 1962) developed a model of the primary production with particular emphasis on the North Sea. The model was later slightly modified and adapted to new data by Lassen and Nielsen (1972) and by Droissart et al. (1973) in a study of the application of hybrid computation to interaction models.

Considering space averages over a box which is identified with the euphotic layer above the thermocline, two state variables are selected:

(1) the concentration of phosphate C,
(2) the concentration of phytoplankton N.

The evolution of the state variables is examined under the action of light (photosynthesis), zooplankton (grazing) and the transfers across the thermocline.

These effects are described in terms of four control parameters:

(1) the surface lighting $\lambda(t)$,
(2) the zooplankton concentration $Z(t)$,
(3) the mixing frequency* $m(t)$,
(4) the sinking frequency* $s(t)$.

The functions λ, Z, m and s are given empirical forms inferred from the observations. For instance, the mixing coefficient (shown in Fig. 4.3) is ap-

---

* "Frequency" denoting a coefficient of dimension t^{-1} with no suggestion of periodicity.

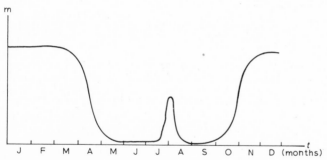

Fig. 4.3. Variation of mixing coefficient over a year (after Lassen and Nielsen, 1972).

proximated by the sum of 3 analytical functions which represent the mixing coefficient in summer, fall and winter respectively.

The evolution equations are written in the form:

$$\frac{dN}{dt} = [p(\lambda)\, f(C) - gZ - s]\, N - m(N - N_0) \tag{4.10}$$

$$\frac{dC}{dt} = -\alpha p(\lambda)\, f(C)\, N - m(C - C_0) \tag{4.11}$$

where N_0 and C_0 denote the values of N and C in the lower layer and where g and α are given constants. The functions $p(\lambda)$ and $f(C)$ represent the effects of light and nutrient limitation on the production rate of the phytoplankton.

In the early version of the model, Steele (1962) suggested that there is a critical value λ_c of λ below which no photosynthesis takes place and that for values of λ greater than λ_c, p is approximately constant (Fig. 4.4).

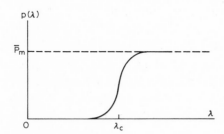

Fig. 4.4. The coefficient of photosynthesis as a function of the surface lighting.

Steele (1962) also introduced a critical value $C_c = 0.6$ matg P/m³ and postulated:

$$f(C) = \begin{cases} \dfrac{C}{C_c} & \text{for} \quad C < C_c \\[2mm] 1 & \text{for} \quad C > C_c \end{cases} \tag{4.12}$$

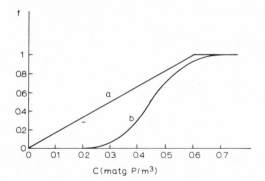

Fig. 4.5 Effect of phosphate concentration on the production of phytoplankton according to Steele (1962) (curve a) and Lassen and Nielsen (1972) (curve b).

Lassen and Nielsen (1972) replaced eq. 4.12 by the continuous curve:

$$f(C) = \frac{1}{\sigma\sqrt{2\pi}} \int_{-\infty}^{c} e^{-\frac{1}{2}\left(\frac{u-a}{\sigma}\right)} du \qquad (4.13)$$

where: $a = 0.45$ and $\sigma = 0.1$. The two curves are shown in Fig. 4.5.

Droissart et al. (1973) attempted to optimize the parameters on a hybrid computer. They found that better results were obtained with eq. 4.13 by taking $a = 0.425$. They also suggested slightly modified expressions for the control parameters.

As an example of typical results, Figs. 4.6 and 4.7 show a comparison between the observed annual variations of the phosphate concentration and the evolution predicted by the model with the assumptions made respectively by Steele (1962), Lassen and Nielsen (1972), Droissart et al. (1973).

One can see that, although the model reproduces the broad trend of the phenomenon, there are significant differences between observed and calculated values. This is partly due to the restriction of the model to only two state variables. One must then either neglect several effects which could be significant or, to account for them, one must multiply the control parameters for which empirical expressions must be substituted in the equations.

We find here exemplified the main shortcoming of all interaction models. They are limited by the quality of the data which they use to identify the interactions, elucidate their laws and determine the empirical expressions of the parameters. (This difficulty will be further discussed by Steele in Chapter 10.)

With Steele's assumptions, the laws of interactions between nutrients, phytoplankton and zooplankton are all linear or bilinear and may be regarded as particular cases of eq. 4.5. When more sophisticated expressions like 4.13 are used, the representation is more elaborate, but one can see that

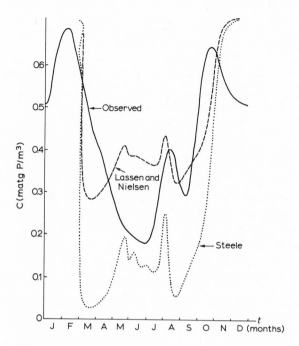

Fig. 4.6. Observed and calculated concentrations of phosphate according to Steele (1962) and Lassen and Nielsen (1972).

eq. 4.5 remains a reasonable approximation and, indeed, that the advantage of the refinements does not show through the results.

In more complicated models with many state variables, one usually finds that a satisfactory representation can be achieved by combining, in sums and products, simple laws like 4.5, 4.6 and 4.7.

For instance, in a situation where both phosphates and nitrates are growth-limiting, one can express the total rate of uptake of nutrients by

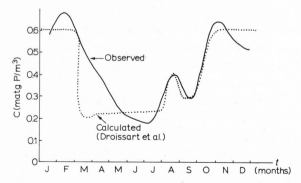

Fig. 4.7. Observed and calculated concentrations of phosphate according to Droissart et al. (1973).

phytoplankton as a product of two Michaelis-Menten-Monod laws (Di Toro et al., 1971).

To give another example, it may be interesting to discuss briefly the model presented by Pichot for the Ostend Bassin de Chasse (Pichot, 1973).

The Bassin de Chasse is a closed sea basin with negligible interactions with the exterior. Considering only box averages over the whole basin, Pichot selects four state variables on the basis of a systematic study of a long series of data:

(1) the nutrients (nitrites and nitrates) concentration C,
(2) the phytoplankton concentration N,
(3) the zooplankton concentration Z,
(4) the bacterial number B.

(The reasons for not considering phosphate in this particular case were explained in section 4.3 as a particular example of eq. 4.9).

Incident light and wind speed are taken as control parameters and several auxiliary variables like temperature, recirculation of sediments, extinction of light, turbulent mixing, etc., are expressed in terms thereof and some of the state variables by algebraic equations inferred from the observations.

The evolution equations are written in the form:

$$\frac{dC}{dt} = S + aB - f_1 CN(1 - \alpha N) \tag{4.14}$$

$$\frac{dN}{dt} = f_2 CN(1 - \alpha N) - k_1 N - f_3 Z(1 - \beta Z) \tag{4.15}$$

$$\frac{dZ}{dt} = f_4 Z(1 - \beta Z) - k_2 Z \tag{4.16}$$

$$\frac{dB}{dt} = f_5 ZB(1 - \gamma B) + k_3 B \tag{4.17}$$

where a, f, k, α, β, ..., are known functions of time; S is a source term and corresponds to the dissolution of sediments; aB represents a production of nutrients by bacteria (nitrification). Linear laws are assumed for the processes of respiration, excretion, death and sedimentation which combine in terms like $-k_1 N$, $-k_2 Z$, etc. Feeding (or nutrient uptake) is represented by the product of a Michaelis-Menten-Monod law and a logistic law. The approximation 4.8 is made for the uptake of nutrient and the feeding of bacteria on zooplankton's excretions while the limiting case of eq. 4.9 is assumed for zooplankton grazing.

Comparison of the model's predictions with the observations (Figs. 4.8 and 4.9) shows once more that, although the general behaviour is the same, significant differences remain in the details.

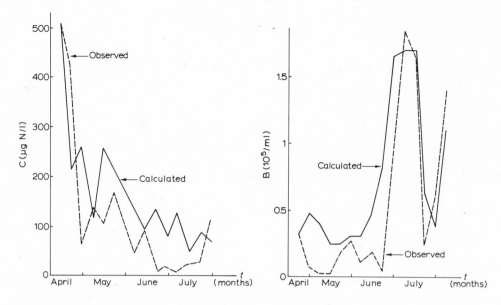

Fig. 4.8. Comparison between observed and calculated concentrations of nutrients in the Ostend Bassin de Chasse (after Pichot, 1973).

Fig. 4.9. Comparison between observed and calculated bacterial population in the Ostend Bassin de Chasse (after Pichot, 1973).

The limitations of the interaction models are again emphasized. One may reasonably believe however that they are, for a large part, due to an excess of simplicity. The severe restriction of the number of state variables, in many cases, leaves in the shadow some important interactions and, in particular, all the physics is concealed in a few control parameters which can only badly describe such essential effects as turbulent mixing or bottom erosion.

The limited success of the present ecological models is a warning and a challenge. Before one can rely on ecological simulation for predictions and management, one certainly needs much more information from observations and laboratory research. But it may be that, no matter the amount of data, simple ecological models cannot work much better. Perhaps it is just not possible to pack so many complex phenomena in a few empirical control functions. Perhaps this is a case where one needs an increased sophistication and one should urge the development of more ambitious models as the general multidisciplinary models which were advocated in the introduction of this volume.

REFERENCES

Di Toro, D.M., O'Connor, D.J. and Thomann, R.V., 1971. *Adv. Chem. Ser.*, 106: 131.

Droissart, A., Leroy, R. and Smitz, J., 1973. *Programme national sur l'environnement physique et biologique, projet mer, 22.* Department of Scientific Policy, Brussels.

Fuhs, G.N., Demmerle, S.D., Canelli, E. and Chen, M., 1972. *Am. Soc. Limnol. Oceanogr., Spec. Symp.*, 1: 113.

Kelly, R.A., 1971. *The Effects of Fluctuating Temperature on the Metabolism of Freshwater Microcosms.* Thesis, University of North Carolina, Durham, N.C.

Kelly, R.A., 1974. In: B.C. Patten (Editor), *Systems Analysis and Simulation in Ecology, 3.* Academic Press, New York, N.Y.

Lassen, H. and Nielsen, P.B., 1972. *Simple Mathematical Model for the Primary Production.* ICES, Plankton Committee CM 1972/L : 6.

Lotka, A.J., 1956. *Elements of Mathematical Biology.* Dover Publ., New York, N.Y., 465 pp.

Odum, H.T., 1972. In: B.C. Patten (Editor), *Systems Analysis and Simulation in Ecology, 1.* Academic Press, New York, N.Y., pp. 139—211.

Parsons, T.R. and Le Brasseur, R.J., 1970. In: J.M. Steele (Editor), *Marine Food Chain.* Oliver and Boyd, Edinburgh, pp. 325—343.

Pichot, G., 1973. Nato Science Committee Conference on Modelling of Marine Systems, Ofir, Portugal, 1973.

Sen Gupta, R., 1969. *Tellus*, 21: 270.

Slobodkin, L.B., 1966. *Growth and Regulation of Animal Populations.* Holt, Rinehart and Winston, New York, N.Y., 192 pp.

Steele, J.H., 1956. *J. Mar. Biol. Assoc. U.K.*, 35: 1.

Steele, J.H., 1958. *Mar. Res. Scot.*, 7: 36.

Steele, J.H., 1959. *Biol. Rev.*, 34: 129.

Steele, J.H., 1962. *Limnol. Oceanogr.*, 7: 137.

CHAPTER 5

NON-LINEARITIES ASSOCIATED WITH PHYSICAL AND BIOCHEMICAL PROCESSES IN THE SEA

Norman S. Heaps and Yves A. Adam

5.1. INTRODUCTION

The purpose of this chapter is to point out the importance of some of the inherent non-linearities present in both physical and biochemical processes in the sea. To this end, we discuss the influence of terms in the relevant mathematical equations which represent second- and higher-order effects.

It is a fact of experience that, according to location and circumstance, the non-linear terms in the dynamical equations for the sea may, at one extreme, produce effects which dominate the sea's motion, or, oppositely, have quite a negligible influence. In the latter case, linearization is justified. However, even when non-linearities are significant, a linear first-order solution may constitute an initial step towards the determination of the complete motion and, moreover, might provide a relatively simple theoretical framework from which to gain a better understanding of the water movements.

Therefore, while wishing to stress the dangers of excessive linearization, it seems prudent at the outset to suggest that the value of linear solutions should not be underestimated.

5.2. PHYSICAL PROCESSES

On the physical side, to illustrate non-linear effects, we shall refer particularly to certain properties of tides and storm surges. Lamb (1932, pp. 280–282) presented the case of tidal waves progressing along a uniform canal of rectangular section, taking the dynamical equations as:

$$\frac{\partial u}{\partial t} + u \frac{\partial u}{\partial x} = -g \frac{\partial \zeta}{\partial x} \tag{5.1}$$

$$\frac{\partial}{\partial x}[(h + \zeta)u] + \frac{\partial \zeta}{\partial t} = 0 \tag{5.2}$$

where t denotes the time, x a coordinate measured along the canal in the direction of wave propagation, u the x-directed current, ζ the tidal elevation and h the mean depth. The waves were considered to originate from a simple harmonic oscillation at $x = 0$ given by

$$\zeta = a \cos \sigma t \tag{5.3}$$

A first approximation to the motion in the canal was derived by solving the linearized forms of 5.1 and 5.2, omitting the non-linear terms $u \partial u/\partial x$ and $\partial(\zeta u)/\partial x$. Waves of symmetrical profile and frequency σ resulted. Subsequently, the linearized solution was perturbed, substitution in eqs. 5.1 and 5.2 then yielding a second approximation to the motion with added expressions representing non-linear effects. These expressions describe a tide with amplitude proportional to a^2 and frequency 2σ, twice that of the primary disturbance. The presence of this tide, induced by the non-linear terms, distorts the wave profile such that, at any fixed position, the fall of surface level occupies a longer time than the rise (e.g. see Fig. 5.1). The latter is a well-known tidal phenomenon in shallow bays and estuaries. Continuing the approximation procedure gives tides of successively higher orders, whose frequencies are 3, 4, 5, ... n times that of the primary. Thereby, the non-linear terms generate an infinite set of tidal harmonics from the original input (eq. 5.3). If the input were to consist of *two* oscillations:

$$\zeta = a \cos \sigma t + a' \cos(\sigma' t + \epsilon) \tag{5.4}$$

the second approximation would yield tides of frequencies $\sigma + \sigma'$ and $\sigma - \sigma'$. Subsequent higher approximations would then produce tides with a range of frequencies obtained by combining integer multiples of σ and σ'. This demonstrates that the effect of the non-linearity is not only to generate tidal harmonics of each primary wave, but also compound harmonics corresponding to linear combinations of the primary-wave frequencies. Clearly, non-linear interaction between the primary tidal waves gives rise to a great complexity of higher-order tides.

While Lamb's problem described above is simple and purely theoretical, it nevertheless serves to illustrate a behaviour which is of considerable practical importance in the analysis and prediction of tides in shallow water. Thus, in tidal rivers and estuaries, harmonics of the basic astronomical tidal constituents develop prominently due to non-linear influences. A condition for such development is that the ratio of tidal elevation to depth (a/h) is not small. Appropriately, harmonics of this type are called shallow-water tides. Their presence greatly complicates tidal prediction and, to account for them in this respect, special methods have had to be evolved as extensions to the application of classical harmonic theory (Doodson, 1957; Rossiter and Lennon, 1968). A notable example of the part played by shallow-water tides in

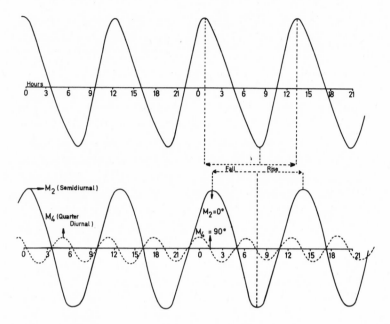

Fig. 5.1. Distorted wave profile due to a superposition of primary and secondary oscillations (in this case, the M_2 and M_4 tides respectively).

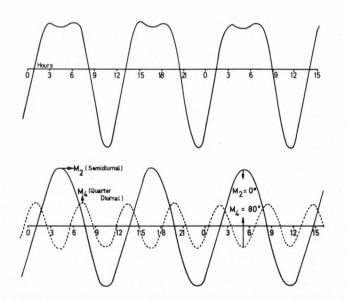

Fig. 5.2. Shallow-water effect giving a combination producing a double high water.

determining the shape of the tide curve at a port is the occurrence of double high waters at Southampton. Doodson and Warburg (1941, pp. 218—226) explain the formation of double high waters in terms of the superposition of quarter, sixth and eighth diurnal tides on a primary semi-diurnal oscillation. A simple illustration of their reasoning, involving the M_2 and M_4 tidal constituents as primary and first harmonic respectively, is given in Fig. 5.2.

Eqs. 5.1 and 5.2 are fundamental to the mathematical theory of surface waves in shallow water, accounting for the formation of breakers and bores (Stoker, 1948). In tides, the equations have been used analytically by several workers to investigate non-linear effects, e.g. Fjeldstad (1941), Proudman (1953, pp. 293—295), but manifestly this work did not consider the influence of turbulent friction. Kreiss (1957) included a frictional term in studying tidal oscillations in a uniform channel: connected with the open sea at one end and closed at the other. As a result of non-linear influence, greatest in conditions of resonance, he was able to explain the asymmetry between flood and ebb currents often observed in a tidal estuary. Normally in tidal dynamics a quadratic law of bottom friction has been assumed, taking the form:

$$F = k\rho u |u| \tag{5.5}$$

for one-dimensional motion in a horizontal x-direction, where F denotes the bottom stress, ρ the density of the water and k a constant coefficient (Proudman, 1953, pp. 136, 303). By its very nature, the quadratic formulation introduces a non-linearity. Linearization of the form is possible for a pure harmonic tide (Proudman, 1953, p. 310), a procedure which constitutes one of the basic elements in the Harmonic Method of Tidal Computation (Dronkers, 1964). With eq. 5.5, the equations governing the propagation of long waves along a narrow estuary may be written:

$$\frac{\partial u}{\partial t} + u \frac{\partial u}{\partial x} = -g \frac{\partial \zeta}{\partial x} - \frac{ku|u|}{h + \zeta} \tag{5.6}$$

$$\frac{\partial}{\partial x} (Au) + b \frac{\partial \zeta}{\partial t} = 0 \tag{5.7}$$

where A denotes the area of cross-section of the estuary and b the breadth of the water surface at any position x and time t. The frictional expression $ku|u|/(h + \zeta)$ frequently becomes the most important non-linear term in shallow water and generates overtides on its own account (Doodson and Proudman, 1923) in addition to those which arise from the product terms $u\partial u/\partial x$, $\partial(Au)/\partial x$ and $b\partial\zeta/\partial t$. The terms $\partial(Au)/\partial x$ and $b\partial\zeta/\partial t$ are second order since, for a cross-sectional periphery of general shape, both A and b vary with ζ. In practice, only numerical solutions of eq. 5.6 and 5.7 can take account of all the non-linearities, and in this respect finite-difference solutions have proved to be invaluable in tidal computations (e.g. Rossiter and Lennon, 1965; Dronkers, 1969). The convective term $u \partial u / \partial x$ usually has quite a small effect but can become of importance in those estuaries where large changes in either b or h occur in comparatively short distances. Also, the term is partly responsible for the commonly observed phenomenon in tidal rivers whereby the mean surface level increases continuously from the mouth to the head (Rossiter, 1967, pp. 1551—1552).

The interaction between tide and storm surge in a shallow estuary is a further non-linear mechanism affecting sea-level variation. Theoretical investigations into this topic have been carried out by Proudman (1955a,b, 1957). Employing basically eqs. 5.6 and 5.7, he obtained analytic solutions, up to the second order, for the propagation of tide and surge in an estuary after generation in the open sea. As a result, it was possible to make some general deductions concerning the manner in which the tide modifies the surge. Thus, for the same meteorological conditions over the sea, and a progressive wave of tide and surge, he found that the height of a surge whose maximum occurs near to the time of tidal high water is less than that of a surge whose maximum occurs near to the time of tidal low water, the height decreasing with increasing range of tide. This provides an explanation of the observation that large surge peaks tend to avoid high water, a fact of considerable importance in studying the conditions of occurrence of abnormally high sea levels. Rossiter (1961) examined specifically interaction in the Thames Estuary, representing the estuary by a channel of constant depth and exponentially varying breadth. Eqs. 5.6 and 5.7 were solved numerically to yield surge elevations along the length of the estuary for different combinations of tide and surge introduced at the mouth. The surge profile was shown to be radically affected by the phase relationship between tide and surge, the main interaction effect taking the form of an amplification of surge height on the rising tide — a behaviour reflected in observational statistics. Rossiter gave a simple physical explanation of this phenomenon in terms of a retardation or hastening of the predicted tide due to the presence of the surge, the latter changing the depth of water and hence the characteristics of tidal wave propagation. In work with a two-dimensional numerical model, Banks (1970) showed that non-linear effects involving tide and surge can be of significance throughout the entire southern North Sea, as well as in the Thames.

Thus far, we have been mainly concerned with the influence of the non-linear terms on sea level, particularly referring to those of its variations associated with the tides. The regime of currents in the sea is also affected by the essential non-linearity of the marine system and in this respect mention is now made of the development, during recent years, of boundary-layer theory predicting steady second-order currents near the sea bed induced by tides and gravity waves. Both in estuaries and in the open sea, such residual currents, when taken in conjunction with the oscillatory motion, evidently play an important part in the movement of bed material, contributing to particle drift velocity, or mass transport, just above the sea floor (Longuet-Higgins, 1953). Theoretical papers by Abbott (1960), Hunt (1961) and Johns (1967, 1968, 1970) have determined bottom currents and the associated net transports brought about by tidal oscillations in estuaries with applications to sediment deposition in the Thames and the Humber. Extensions of the work to the open-sea situation have been carried out by Hunt and Johns (1963) and Johns and Dyke (1971, 1972), the latter investigation presenting calculations relevant to conditions in Liverpool Bay. Further work, with numerical models, is required in this field in order to apply theoretical concepts already established to real sea basins.

Having reviewed in the preceding account a number of non-linear effects in tidal dynamics, referring particularly for purposes of simple mathematical illustration to the one-dimensional propagation of long waves along a narrow channel, it seems appropriate at this stage to take the discussion to a more general level by examining certain non-linear features of the three-dimensional equations of motion of the sea. Eq. 1.10 can be written:

$$\frac{\partial u}{\partial t} + 2\,\Omega \wedge u + \nabla \cdot u\,u = -\nabla\left(\frac{p}{\rho_{\mathrm{m}}}\right) + G + \mathcal{D} \tag{5.8}$$

where G is the mean body force; its horizontal components represent the tidal influence, its vertical component reduces to gravity; u defines the mean motion which, incremented by the turbulent fluctuations w, yields the total field of flow; p is the pressure; Ω is the rotation vector of the earth; ρ is the reference density.

The last term in the right-hand side is given by eq. 1.11:

$$\mathcal{D} = -\nabla \cdot \langle ww \rangle + \frac{\mu}{\rho_{\mathrm{m}}}\,\nabla^2 u \tag{5.9}$$

where μ is the dynamic molecular viscosity and where an angular bracket represents an average taken over some fundamental interval.

The tensor:

$$\langle ww \rangle \tag{5.10}$$

may be interpreted as a system of internal stresses existing because of turbulence. Because they involve products of the turbulent velocities, these stresses may be regarded as the outcome of a non-linear process. Clearly, therefore, $\mathcal{D}$ in eq. 5.8 is non-linear and importantly so from the research point of view, since its origin is in turbulence — a phenomenon which poses a major problem in fluid dynamics (Stewart, 1967). It is commonplace to relate the stresses to space gradients of the mean motion using coefficients of eddy viscosity. Typically (eq. 1.13):

$$\mathcal{D}_j = \frac{\partial}{\partial x_1}\left(\nu_1 \frac{\partial u_j}{\partial x_1}\right) + \frac{\partial}{\partial x_2}\left(\nu_2 \frac{\partial u_j}{\partial x_2}\right) + \frac{\partial}{\partial x_3}\left(\nu_3 \frac{\partial u_j}{\partial x_3}\right) \qquad (5.11)$$

Eddy coefficients such as ν_1, ν_2 and ν_3 depend on the motion but, for convenience, have often been regarded as time independent and simply distributed. Their validity in respect of giving a correct representation of basic physical processes has been questioned by Stewart (1956).

Thus, the hydrodynamical eq. 5.8 contains non-linearities arising from averages of products of the motion, as in 5.10. The equations of mass balance involve similar non-linear terms. For example, the rate of transport of salt per unit area perpendicular to the x_1-axis, averaged over a fundamental interval, is (using the notations of Table 1.1):

$$\langle \rho_s v_1 \rangle = r_s u_1 + \langle w_1(\rho_s - r_s) \rangle \qquad (5.12)$$

where r_s and $(\rho_s - r_s)$ denote the mean and turbulent salinities. Here $r_s u_1$ represents the rate of transport associated with the mean motion and $\langle w_1(\rho_s - r_s) \rangle$ that associated with turbulent diffusion. Both terms are non-linear and contribute to the equation of salt conservation (Proudman, 1953, p. 89). An eddy-diffusion coefficient κ_1 may be defined such that:

$$\langle w_1(\rho_s - r_s) \rangle = -\kappa \frac{\partial r_s}{\partial x_1} \qquad (5.13)$$

Further, consider the rate of transport of salt, Q, across a vertical section of unit width perpendicular to the x_1-axis, extending from the sea surface to the sea bed. We have (cf. eq. 3.59):

$$Q = \int_{-h}^{\zeta} [r_s u_1 + \langle w_1(\rho_s - r_s) \rangle]\, dx_3$$

$$= H(\overline{r_s u_1}) + H\langle \overline{w_1(\rho_s - r_s)} \rangle$$

where

$$H = h + \zeta$$

and where bars denote depth-averaged values. Writing:

$$u_1 = \bar{u}_1 + \hat{u}_1$$

$$r_s = \bar{r}_s + \hat{r}_s$$

it follows that:

$$Q = H \, \bar{u}_1 \, \bar{r}_s + H \, \overline{\hat{r}_s \, \hat{u}_1} + H \langle \overline{w_1 (\rho_s - r_s)} \rangle \tag{5.14}$$

Here, the first term represents salt transport by the depth-mean flow and the final term salt transport by turbulent diffusion. The middle term is non-zero when there is a correlation along the vertical between $\hat{u}_1$ and $\hat{r}_s$, the deviations at a particular depth from the depth-mean values $\bar{u}_1$ and $\bar{r}_s$. Such a correlation occurs in a shearing current when a vertical gradient of horizontal velocity exists with vertical turbulent mixing (Bowden, 1963, 1965). In these circumstances, the process of horizontal salt transport represented by $H\overline{\hat{r}_s \hat{u}_1}$ exemplifies the so-called "shear effect". This effect may be associated with wind-driven, density or tidal currents and, in the latter case, is known to have an important influence in the dispersion of pollutants in the Southern Bight of the North Sea (Nihoul, 1972). All three terms in eq. 5.14 are non-linear and contribute to the vertically integrated form of the equation of salt conservation.

In eq. 5.14, the first term in the right-hand side may be associated with the advection of salt by the bulk flow and the last two terms with the diffusion of salt by shear effect and turbulence, respectively.

A final reference to eq. 5.8 concerns the non-linear term $\nabla \cdot uu$. This gives the convective part of the acceleration of a fluid element. In tidal motion the terms generally have a small effect, only becoming significant in shallow coastal and estuarial waters where they generate steady surface displacements and tidal harmonics of even order. Brettschneider (1967), in a numerical analysis of North Sea tides, writes:

$$u_1 \frac{\partial u_1}{\partial x_1} + u_2 \frac{\partial u_1}{\partial x_2} + u_3 \frac{\partial u_1}{\partial x_3} = -(u_1^2 + u_2^2) \frac{\partial}{\partial x_2} \tan^{-1}\left(\frac{u_2}{u_1}\right)$$

$$-u_1 \frac{\partial u_3}{\partial x_3} + u_3 \frac{\partial u_1}{\partial x_3} \tag{5.15}$$

The first term on the right-hand side of 5.15 predominates (due to essentially horizontal flow) and demonstrates that the convective acceleration is greatest when horizontal current is high and altering direction with changing position. However, such conditions are generally restricted to a few specific locations determined mainly by coastal configuration and depth variation (e.g. see Crean, 1972).

Within the limited scope and purpose of this note, it is scarcely possible to extend the discussion to cover all the known and studied non-linear physical processes in the sea. The entire marine system is, by its very nature, non-linear. An important and expanding area of research, not mentioned, which perhaps more than any other illustrates the complexity and pervasive character of the non-linear processes, is that dealing with the properties of surface and internal waves in the open ocean and near coastal boundaries. Surveys of this subject have been given by Phillips (1966) and Stoker (1957). Further, there have been developments in the theory of ocean circulation involving non-linear dynamical equations, e.g. Stommel (1958), Bryan (1963), Bryan and Cox (1968a,b).

5.3. BIOCHEMICAL AND ECOLOGICAL PROCESSES

If we now turn to biochemical and ecological interactions in the box model approximation, the system is described by a set of differential equations of the form (cf. eq. 4.3):

$$\frac{d\bar{r}_\alpha}{dt} = \bar{I}_\alpha \qquad \alpha = 1, ..., n \tag{5.16}$$

where, external inputs not being considered, the right-hand side reduces to the interaction term $\bar{I}_\alpha$.

As mentioned in Chapter 4, the simplest possible interaction law is the linear one, i.e. using eq. 4.4:

$$\frac{d\bar{r}_\alpha}{dt} = k_{\alpha\beta}\,\bar{r}_\beta \tag{5.17}$$

It is easy to show that a completely linear system of equations of the form 5.17 is a very crude approximation indeed in biological and chemical modelling. Let us consider a simple model of primary production limited to nutrient—phytoplankton interactions of the form:

$$\frac{dx_1}{dt} = kx_1 \tag{5.18}$$

$$\frac{dx_2}{dt} = -kx_1 \tag{5.19}$$

where x_1 is the concentration of nutritive elements in the phytoplankton, x_2 their concentration in the sea water and where it has been assumed that the total mass of these elements is conserved, since:

$$\frac{d(x_1 + x_2)}{dt} = 0 \tag{5.20}$$

The solution is shown in Fig. 5.3 (curve a): x_1 grows exponentially with-

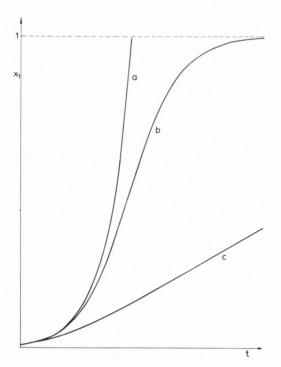

Fig. 5.3. Concentration of nutritive elements in phytoplankton calculated with (*a*) a linear law, (*b*) a logistic law, and (*c*) a Michaelis-Menton-Monod law.

out any limit, even when the nutrient supply is completely exhausted. The amount of nutrient in the water can become negative after some time, which is clearly a nonsense.

Let us now examine the effect of non-linearities and modify eqs. 5.18 and 5.19 as follows:

$$\frac{dx_1}{dt} = \alpha x_1 x_2 \tag{5.21}$$

$$\frac{dx_2}{dt} = -\alpha x_1 x_2 \tag{5.22}$$

It is easily seen that now exponential unbounded growth and negative nutrient concentrations are no longer possible because when x_2 tends to zero, the rates dx_1/dt and dx_2/dt both tend to zero.

In fact eq. 5.20 still holds, and eliminating x_2, eq. 5.21 reads:

$$\frac{dx_1}{dt} = \alpha C x_1 - \alpha x_1^2 \tag{5.23}$$

where:

$$C = x_1 + x_2$$

We thus obtain the logistic equation 4.6. The solution is shown on Fig. 5.3 (curve b).

As a final example, let us consider the Michaelis-Menten-Monod law and replace eqs. 5.18 and 5.19 by:

$$\frac{dx_1}{dt} = \beta \frac{x_1}{b_1 + x_1} \frac{x_2}{b_2 + x_2} \tag{5.24}$$

$$\frac{dx_2}{dt} = -\beta \frac{x_1}{b_1 + x_1} \frac{x_2}{b_2 + x_2} \tag{5.25}$$

The solution is shown on Fig. 5.3 (curve c). A steady state is reached ultimately (but it does not appear on the diagram).

As a final example, we now discuss briefly the chemical model proposed by Adam (1973) for the different forms of phosphorus in the Ostend Bassin de Chasse.

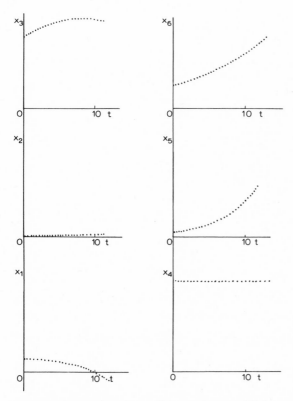

Fig. 5.4. Evolution of six different forms of P in a closed sea basin assuming completely linear interactions (after Adam, 1973).

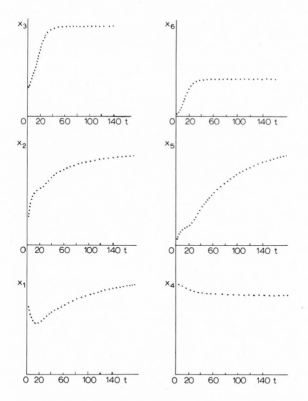

Fig. 5.5. Evolution of six different forms of P in a closed sea basin assuming quadratic-bilinear interactions (after Adam, 1973).

Six state variables were considered representing the total amounts of phosphorus in six compartments:

(1) dissolved phosphates in free water   : x_1
(2) non-living matter in suspension   : x_2
(3) dissolved phosphates in interstitial water : x_3
(4) bottom sediments   : x_4
(5) plankton   : x_5
(6) benthos   : x_6

The interaction rates are expressed in terms of sums and products of bilinear quadratic and Michaelis-Menten-Monod laws. Figs. 5.4, 5.5 and 5.6 show the results of the simulation in three typical cases: (1) Fig. 5.4 — completely linear case; (2) Fig. 5.5 — partially linear case, the linearization is limited to the Michaelis-Menten-Monod expressions (cf. eq. 4.8); and (3) Fig. 5.6 — strongly non-linear case.

The unit of time is a day. The shortcomings of the completely linearized equations are evident: unlimited growths, negative concentrations, etc. Com-

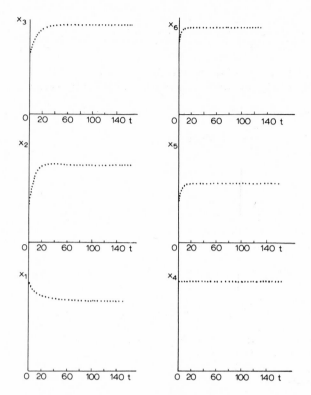

Fig. 5.6. Evolution of six different forms of P in a closed sea basin assuming strongly non-linear interactions (after Adam, 1973).

paring the two non-linear systems, one can see that the time variations are more pronounced in the quadratic-bilinear case.

In the strongly non-linear case, a steady state is rapidly approached.

A more detailed description and further results can be found in Adam (1973).

REFERENCES

Abbott, M.R., 1960. *J. Mar. Res.*, 18: 83.
Adam, Y.A., 1973. Nato Science Committee Conference on Modelling of Marine Systems, Ofir, Portugal, 1973.
Banks, J.E., 1970. *A Mathematical Model Investigation of Tide, Surge and Interaction in the Thames—Southern North Sea Region.* Thesis, University of Liverpool, Liverpool.
Bowden, K.F., 1963. *Int. J. Air Water Pollut.*, 7: 343.
Bowden, K.F., 1965. *J. Fluid Mech.*, 21: 83.
Brettschneider, G., 1967. *Mitt. Inst. Meereskd. Univ. Hamburg*, 7.
Bryan, K., 1963. *J. Atmos. Sci.*, 20: 594.
Bryan, K. and Cox, M.D., 1968a. *J. Atmos. Sci.*, 25: 945.

Bryan, K. and Cox, M.D., 1968b. *J. Atmos. Sci.*, 25: 968.

Crean, P.B., 1972. *Numerical Model Studies of the Tides Between Vancouver Island and the Mainland*. Thesis, University of Liverpool, Liverpool.

Doodson, A.T., 1957. *Int. Hydrogr. Rev.*, 34 (1): 85.

Doodson, A.T. and Proudman, J., 1923. *Rep. Br. Assoc. Adv. Sci.*, 1.

Doodson, A.T. and Warburg, H.D., 1941. *Admiralty Manual of Tides*. Her Majesty's Stationery Office, London.

Dronkers, J.J., 1964. *Tidal Computations in Rivers and Coastal Waters*. North-Holland Publishing Company, Amsterdam, 530 pp.

Dronkers, J.J., 1969. *J. Hydraul. Div. Am. Soc. Civ. Engr.*, 95: 29.

Fjeldstad, J.E., 1941. *Astrophys. Norv.*, 3: 223.

Hunt, J.N., 1961. *Tellus*, 13: 79.

Hunt, J.N. and Johns, B., 1963. *Tellus*, 15: 343.

Johns, B., 1967. *Geophys. J.R. Astr. Soc.*, 13: 377.

Johns, B., 1968. *Geophys. J.R. Astr. Soc.*, 15: 501.

Johns, B., 1970. *Geophys. J.R. Astr. Soc.*, 20: 159.

Johns, B. and Dyke, P., 1971. *Geophys. J.R. Astr. Soc.*, 23: 287.

Johns, B. and Dyke, P., 1972. *J. Phys. Oceanogr.*, 2: 73.

Kreiss, H., 1957. *Tellus*, 9: 53.

Lamb, H., 1932. *Hydrodynamics*. Cambridge University Press, London, pp. 280—282.

Longuet-Higgins, M.S., 1953. *Philos. Trans. R. Soc. Lond.*, A 245: 535.

Nihoul, J.C.J., 1972. *Bull. Soc. R. Sci. Liège*, 10: 521.

Phillips, O.M., 1966. *The Dynamics of the Upper Ocean*. Cambridge University Press, London, 262 pp.

Proudman, J., 1953. *Dynamical Oceanography*. Methuen & Co., London, 421 pp.

Proudman, J., 1955a. *Proc. R. Soc. Lond.*, 231A: 8.

Proudman, J., 1955b. *Proc. R. Soc. Lond.*, 233A: 407.

Proudman, J., 1957. *J. Fluid Mech.*, 2: 371.

Proudman, J., 1958. *J. Fluid Mech.*, 3: 411.

Rossiter, J.R., 1961. *Geophys. J.R. Astr. Soc.*, 6: 29.

Rossiter, J.R., 1967. In: S.K. Runcorn (Editor), *International Dictionary of Geophysics*, 2. Pergamon Press, Oxford.

Rossiter, J.R. and Lennon, G.W., 1965. *Proc. Inst. Civ. Engr.*, 31: 25.

Rossiter, J.R. and Lennon, G.W., 1968. *Geophys. J.R. Astr. Soc.*, 16: 275.

Stewart, R.W., 1956. *Can. J. Phys.*, 34: 722.

Stewart, R.W., 1967. In: *Proc. Canadian Congress of Applied Mathematics, Quebec*, 3. *General Lectures*, pp. 145—162.

Stoker, J.J., 1948. *Comm. Pure Appl. Math.*, 1: 1.

Stoker, J.J., 1957. *Water Waves*. Interscience Publishers Inc., New York, N.Y., 595 pp.

Stommel, H., 1958. *The Gulf Stream*. University of California Press, Berkely, Calif., 202 pp.

CHAPTER 6

MARINE ECOSYSTEMS WITH ENERGY CIRCUIT DIAGRAMS

Howard T. Odum

6.1. INTRODUCTION

If the parts and processes of marine systems develop a unity of organization of physical, chemical, biological, and geological components to maximize the competitive use of energies available, an overview of these relationships is required for modelling and understanding. Separate modelling by components or by knowledge discipline may not suffice. Justification for these assertions were given in book form (Odum, 1971). If oceanic and atmospheric systems develop an interdependent order that maximizes power because of natural selection, the resulting pattern may involve interplay of all kinds of energies. The magnitudes of potential energy per area per year are as great in biological and chemical phenomena as in physical and geological processes.

Understanding the structure and function of marine ecosystems may be aided by diagramming and evaluating overall models with energy circuit language. This is a visual mathematics which uses energy as a common denominator for combining the models of chemical, physical, biologic, geologic, and economic subsystems. Diagrams help the comparative analysis of model configurations and are a useful interface between the mental concepts and computer simulation. In this review common elements of marine biogeochemical models are diagrammed and expanded in scope to include physical oceanographic terms. The methodology of defining macroscopic mini-models may encompass larger marine systems without increasing detail. Such models emphasize the structures and unity of organization that all systems build for maximizing their competitive survival through better processing. This is opposite from the modelling that fragments into a grid of nodes losing sight of the form and identities the system builds in itself. This chapter considers some uses of the energy language in portraying systems and aiding modelling.

6.2. USES OF ENERGY CIRCUIT LANGUAGE

First consider some examples of recent modelling as they look when

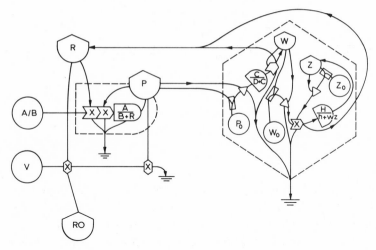

Fig. 6.1. Model for a plankton system translated into energese (from Steele, 1972).

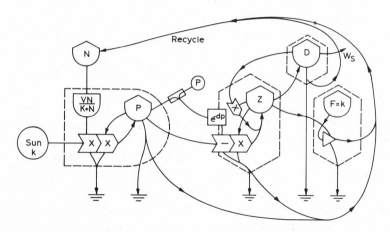

Fig. 6.2. Model for a plankton system translated into energese (from O'Brien and Wroblewski, 1973).

translated into energy circuit language. Figs. 6.1 and 6.2 are examples of a marine ecosystem model, one by Steele (1972) and one by O'Brien and Wroblewski (1973). The differential equations of these authors are exactly translated into the diagrammatic presentation in which each symbol has energetic, mathematic, and kinetic definition. An abbreviated list of symbols is given in Fig. 6.3. More complete discussions, premises, and conventions were given previously (Odum, 1971, 1972, 1973). In this review usual types of mathematical terms are written on the pathways of the diagrams, although this is an unnecessary redundancy, since the language already indicates these by the definitions of the symbols. Selected for this review are

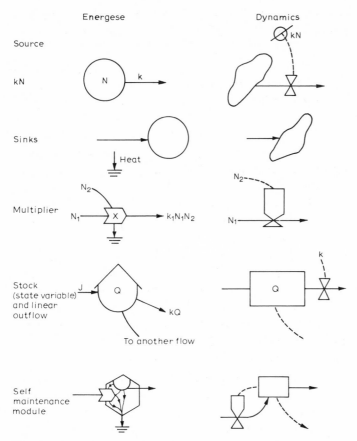

Fig. 6.3. Some main symbols of the energy circuit language compared with Forrester symbols.

discussions of those modules and configurations which were used in the NATO Conference on Modelling of the Sea at Ofir, Portugal.

The following uses of energy language are suggested for modelling the sea:

(1) Analyzing models for comparative consideration of mechanisms. Some mechanisms in one field are recognized by those in another as familiar when they translate.

(2) Combining kinetics, dynamics, energetics, material balance, and economics in one method rather than writing separate equations.

(3) Combining the varied languages of several fields into one common language.

(4) Helping to insure that constraints of energy laws are included. This is done in the calculation of energy balance at each intersection or pathway.

(5) Providing a ready way to recognize emergent mechanisms resulting from combining units. For example, Michaelis-Menten effects emerge whenever a limited flow enters a multiplier.

(6) Providing a recognizable translation between the formulations of mathematics and science and the systems diagrams of Forrester, Koenig, and Payntor.

(7) Presenting differential, difference, logic, and integral equations in a form more readily carried in the mind in single unified form.

(8) Showing complex interactions for purposes of summarizing the impact of proposed environmental actions.

(9) Providing a ready way to portray average data in steady-state flows for purposes of computing missing data. Numbers placed on diagrams indicate visually the magnitudes of time constants, relative importance of flows, and the basis of coefficients in data.

(10) Providing the pathways for which total energy flows are calculated for energy cost benefit evaluations. The relative value of a pathway is taken to be the energy flow which it causes.

(11) When diagrams are provided with computer names and addresses (analog or digital), the diagrams facilitate debugging of simulation programs.

(12) Programming of complex systems in great detail may be followed by successive redrawing with compartmentalization (aggregation) to develop simple models with some of the essence of all classes of components, while retaining the overall integrity of material and energy balance.

(13) The diagrams help to prevent the unintended double insertion of a mathematical term. For example, a Michaelis-Menten relationship may be desired in the model and it is added. Yet the network into which it is added may already have this relationship in the configuration of limited flows reaching multipliers. Thus the modeller adds it twice unknowingly. There may be no harm in this if the system has two. However, hardware is economized and cost is saved by not duplicating unnecessarily those units that are at a micro-level of organization, where inclusion will not change performances much.

(14) Comparison of a great variety of marine ecosystems suggests a pattern of structure, function, and processes common to all systems that endure. The pattern is, as in Fig. 6.4A, one of obligatory storage and development of high-quality energy and information which is fed back to a controlling, pumping improvement of the energy-gathering processes.

Given in Fig. 6.4B is a very simple generalized structure for overall models of marine ecosystems which have some of each kind of subsystem. It is an elaboration of the "reward loop" pattern in Fig. 6.4A. The total generation of power flow is maximized by generating structure, energy, storages, information, and culture from which are fed back forces and amplifiers to pump in the maximum possible energy using all of the potential energy gained towards this. The criterion of competitive continuation and survival is the effective use of structures towards maintaining the energy basis.

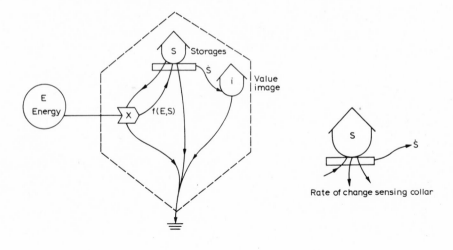

A

B

Fig. 6.4. General form of surviving systems with reward-loop feedback required for Lotka's principle of maximizing power with feedback from stored information, structure, and storages. A. Concept. B. Compartmentalized by physical, chemical, biologic, geologic, and economic sections. Value image i is generated in human interfaces as the sum of the rates of change of all energy storages, S.

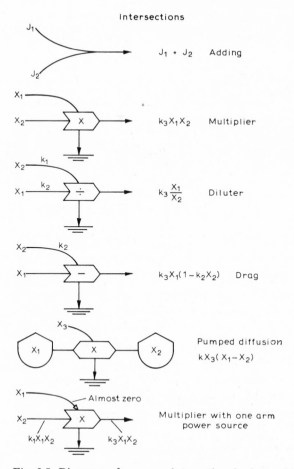

Intersections

$J_1 + J_2$ Adding

$k_3 X_1 X_2$ Multiplier

$k_3 \dfrac{X_1}{X_2}$ Diluter

$k_3 X_1 (1 - k_2 X_2)$ Drag

Pumped diffusion
$k X_3 (X_1 - X_2)$

Multiplier with one arm
power source

Fig. 6.5. Diagrams of common interactional relationships in energy circuit language.

(15) Diagramming as a common language may prevent the unnecessary duplicate development of the same concepts in different fields. For example, the kinetics of solar cells, chloroplasts, photochemical reactions, and some examples of diminishing returns in economics are all identical with the original enzyme substrate recycling model of Michaelis Menten in 1913. (See Fig. 6.5.)

(16) Diagramming in network form shows that many models, which have similar form in algebra or differential equation form, are entirely different phenomena, and this shows in network drawing. For example, many workers in biological oceanography call the limiting-factor action of external nutrients Michaelis-Menten, although the action is not a recycle. Monod, Rashevsky, and Lineweaver-Burke and Kira are the alternative antecedents.

(17) Diagramming with a group gathered around a blackboard has been

found useful in drawing out from several persons their combined knowledge.

(18) Diagramming of systems that are the focus of semantic argument often clarifies issues and meaning. The diagrams of limiting factor kinetics given in this chapter may be an example.

(19) For learning and teaching the diagrams may help in visualizing the various relationships in the environment. For example, chemists and biologists used to working with models in energy circuit language may learn physical oceanography without the mental divisions that comes from changing languages.

6.3. INTERACTION KINETICS

Because much attention has been given to the interactions of nutrients and light in photosynthesis and because there was focus on alternative models for this at the conference, we diagram the alternatives attempting to show the several ways of simplifying the real world system. The discussion considers various uses of the multiplier module related to limiting factors. Included are those equations given by Dugdale in his conference review of some recent approaches. See also previous presentations (Odum, 1971, 1972).

In Fig. 6.6 we remind the reader that chemical reactions or other interactions that require amounts of two entities, flow to the vicinity of the reaction to supply effective concentrations. If the two reactants are supplied with sources that hold concentrations constant, then reactions go as the product. Examples may be found in physics, chemistry, biology, geology, economics and many other fields.

In Fig. 6.6C we let one source (S) remain unlimited, but supply the second (N) at a constant rate of flow to the vicinity of the reaction. Now an increase of S will produce a limiting-factor hyperbola such as was so much discussed by Kira et al. (Shinozaki and Kira, 1956). If a steady state is achieved, a formula is obtained as given by Monod (1942) which looks like the one developed by Michaelis Menten. Many variations result when there are other inflows and outflows to compartment N.

If both sources are inflowing at constant flow, more complex formulae result as given by Rashevsky (see previous discussion, Odum, 1972).

Fig. 6.6B shows the enzyme substrate reaction of Michaelis Menten which also involves a limit to inflow on one side of the multiplier, but here it is limited by recycle because there is a constant quantity of recycling, reused reactant. This network is entirely different from the external limiting pattern in Fig. 6.6C, but at steady state the formula is quite similar. Notice that the system involves a multiplier, plus two sites of storage. The diagramming may

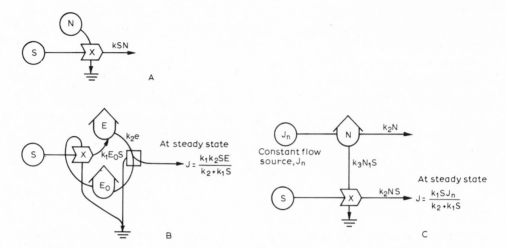

Fig. 6.6. Contrast of internal and external limiting-factor actions both of which have multipliers interactions. A. Unlimited supply. B. Internal recycle limit. C. External supply limit.

help those using hyperbolic equations to realize that their models can be used for the whole ecosystem when materials are recycling and constrained as in microcosms. In that instance the recycling material is nutrients, and the source of main energy is light.

The model has also been used for the initial chlorophyll reaction (Lumry and Rieske, 1959) and for inorganic solar cells (Billig and Plessner, 1949). Here the recycling material is a change in semiconductor state from receptive state to activated one that has charged holes with electrons eliminated.

Since an ecosystem has mineral cycles, contains hundreds of Michaelis-Menten enzyme loops, and starts with the clorophyll action, the ecosystem has Michaelis-Menten loops within loops within loops plus limiting-factor actions of the type in Fig. 6.6B, whenever there are outside sources limiting.

The question facing the modeller is "How many shall I add when only one may be enough to give the model a limiting-factor-type response?" Another question is "Do I recognize a limiting-factor action in my model or do I add a mathematical term, thus adding it twice without knowing it?" If an investigator has phytoplankton in bottles for measurement of limiting-factor kinetics, to how many nested limited reactions is he addressing his measurement efforts? See Fig. 6.7 for a system that has one each of four limiting-factor components in plants.

Another special limiting-factor action is the situation in some oligotrophic phytoplankton which have nutrient pumps that work to collect nutrients actively with feedback from burdened energy reserves. Here one has a Michaelis-Menten internal cycle-pumping inflow from an external compartment which may be externally limited as well.

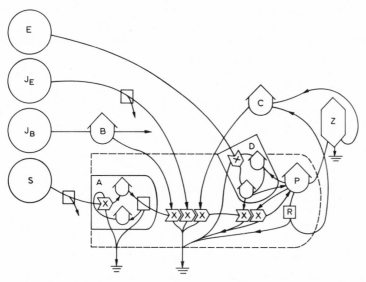

Fig. 6.7. A model of photosynthetic production that has five linked limiting factors operating simultaneously. A = chloroplast with internal recycle limit; B = external limit supply to nutrient B; C = recycle limits in external waters through larger ecosystem; D = limiting action of dark-binding membrane pumps found in some oligotrophic phytoplankton; E = externally limited supply.

Whereas a model with a single recycle arrangement for simulating limiting factors may serve to simulate the nested complexity of many, the coefficients can hardly be identified with the coefficients of any one of the many limiting-factor systems that could be isolated in physiological work. Here one may defeat the purpose of ecological modelling if one tries to identify physiological coefficients with that at the more complex ecological level of aggregated organization.

Consider two ways of getting coefficients after a model is decided upon. In one the organisms are isolated and experiments run on the effects of the interacting inputs. Enclosure changes the system so that the values are not those of the system. Another way is to take observations in the field of the interacting variables and devise ways of estimating rates *in ecos* (in the operating ecosystem). The coefficient is then the only unknown in the equation for that interaction.

Another limiting action which is diagrammed frequently, is the situation when there is a pumped (multiplicative) drain pulling on a flow that is independently determined. In Fig. 6.8 this situation shows the pumped flow taking part of the flow, capable of pulling up to, but not quite all of, the inflow. There are no appreciable storages in this action so that there is always a limiting hyperbolic kinetics. Actually no force may be delivered without a small storage, and the action of a flow in promoting another flow

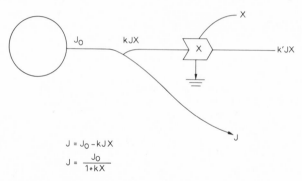

$$J = J_0 - kJX$$

$$J = \frac{J_0}{1+kX}$$

Fig. 6.8. Limiting-factor situation when a multiplier demand, JX, operates from a controlled flow rate, J_0, and local storages are negligible.

does have small storages, but the diagram and algebra in Fig. 6.8 is for the common instance that the storage is too tiny to affect rate responses in the time scale of interest.

Another mechanism being used in some marine models is a multiplicative demand in proportion to that concentration in excess of a threshold with no negative flow. An energy circuit translation of the algebraic expressions is given in Fig. 6.9A and 6.9B. Again the question may be raised whether to add a limit feature or let the system generate its own. Fig. 6.9C has no special threshold feature but will yield no net increase, except when energy input is enough to pay for energy costs of maintaining storage (k_3F). Adding the threshold mechanism over and beyond the inherent minimum for growth (Fig. 6.9D) might be appropriate if the organisms (2) had a behavior mechanism for stopping the feeding work and if the fine detail of the performance were desired.

Another algebraic expression used to indicate a limiting action on an inflow is the exponential expression diagrammed in Fig. 6.9E referred to in some papers as Ivlev's equation (1961), or where light is concerned, the Mitscherlich equation (Verduin, 1964). Whereas the shape of the effect of increasing action on flow is asymptotic, resembling the hyperbolic actions, it differs in being independent of limiting flows (recycle or outside limiting type). The asymptotic property is not subject to the other variables. What the exponential means in terms of mechanism may not be clearly stated.

6.4. PHYSICAL COMPONENTS IN ENERGY LANGUAGE

Whether diagrammed for explanation purposes or for macroview simulation, marine models would be very incomplete without the main physical flows, forces, and storages. In Fig. 6.10—6.17 an attempt is made to translate

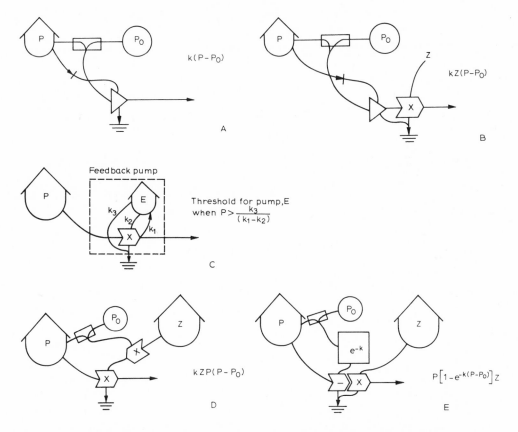

Fig. 6.9. Configurations of energy intake which have thresholds below which there is no growth of rate (see also Figs. 6.6—6.8). A. Linear supply. B. Threshold active demand (multiplier). C. Pseudo-logistic growth; threshold for energy pumping for achieving net growth. D. Threshold control of energy pumping. E. Exponential drag.

the main modules and configurations of the equations of motion into energy languages, the pathways being lines of force balance.

Characteristic of these components are pressure-generating energy storages, energies stored as kinetic energy of velocities, and inertial forces whenever there are accelerations. A storage of a fluid has a pressure in proportion to its quantity Q and its packaging coefficient C (see Fig. 6.10). A force from a pressure potential may generate a velocity and transfer kinetic energy to the motion which may be rotational or translational. Kinetic to potential energy changes and back do not require heat sinks. Kinetic energy is stored relative to the coordinates of the model. When translational velocity exists it has a direction that must be indicated on the diagram. If absent, its direction is taken as into the paper, perpendicular to pathways shown on the paper.

In Fig. 6.10A is shown an arrangement that generates a pressure gradient

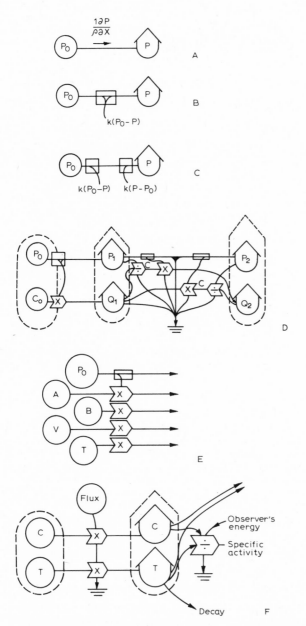

Fig. 6.10. Flows of several types driven by pressure gradient force which is transporting other quantities by advection. A. Pressure gradient force from an outside forcing function exceeding backforce from internal storage. B. Pressure difference sensor; force in proportion to flux-deriving energy from the flux. C. Two flux sensors, one sensing flow from left and one from right, both flows are proportional to pressure gradient. D. Pumped transport of a chemical by pressure gradients. Note state variable is total quantity Q, and ratio to fluid quantity P is concentration. E. One-way advection; transport of chemical quantities, momentum, and heat as in a river inflow. F. Radioactive traces with its carrier.

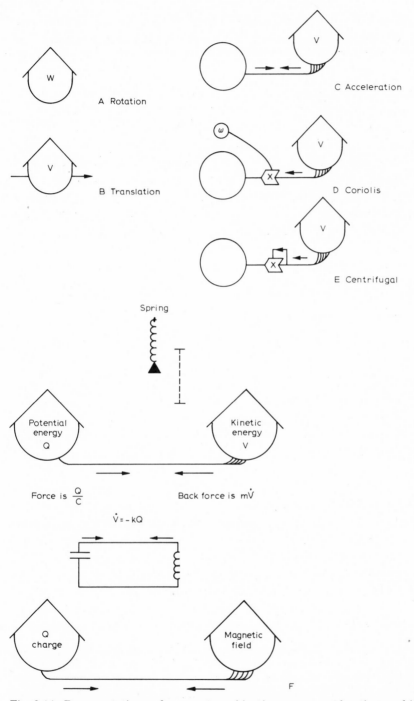

Fig. 6.11. Representations of momentum, kinetic energy, acceleration, and inertial back-force. A. Storage of rotary. B. Translational momentum with accompanying kinetic energies; note vector indication of translational momentum. C. Symbols for acceleration and inertial backforce. D. Symbol for Coriolis-acceleration inertial backforce. E. Symbol for centrifugal inertial backforce. F. Oscillator with potential and kinetic energy.

force according to the difference in pressure. Direction from outside is indicated in Fig. 6.10B as from a river. The flow of fluid is proportional to the
difference in pressure. If the advection is tidal, flow is possible in either
direction and in that case no barb is used for the pathway. In proportion to
the flow of fluid, constituents are transferred as the product of concentration and flux with a string of four multipliers that draw their transport from
the fluid. Transport includes chemical stuffs A and B, momentum, and heat.

In Fig. 6.11A, kinetic energy is indicated by storage units in which the
state variable that carries the energy is either velocity or angular velocity.

Fig. 6.11C shows a convention used where acceleration is generating new
kinetic energy against inertial forces that oppose any acceleration according
to Newtonian convention. Note the brush-like connection of pathway to
tank chosen to indicate inertial energy transfer without heat sink.

Another kind of inertial force for those turning with the earth is that due
to rotation of the earth's axis under a fluid velocity. The product of the earth's
rotation and the velocity is indicated in Fig. 6.11D.

6.5. INERTIAL DISPLACEMENT OSCILLATOR

Oscillators are a favorite class of systems for studies and instruction in
electronics, mechanics, and systems modelling. Vibrating springs, pendulums,
oscillating seiches, sine waves, etc., are various examples. Given in
Fig. 6.11E, in differential equation form and in energy circuit language, is
the essence of a simple oscillator without dissipational elements. Acceleration is negative to the displacement according to one equation form of this.
In energy terms there is a balance of inertial force and potential derived
static force. The inertial backforce is proportional to acceleration and the
driving force from the potential storage is proportional to the storage of the
state variable that goes with displacement. Potential energy contained may
be water or mass of the spring stored against gravity. Another example is the
oscillation between electrical charge and backforce generated by surges of
electric current that induce and store energy as magnetic field.

Molecular diffusion is indicated in Fig. 6.12, with energy for motion
derived from the energy storages within the gradients, the energy emerging as
dispersed heat in the heat sink. If there is a temperature and a chemical
gradient, the two diffusions are coupled according to the Onsager theory as
shown in Fig. 6.12B. In Fig. 6.12C parallel velocity flows exchange momentum through the transfer with molecules diffusing laterally.

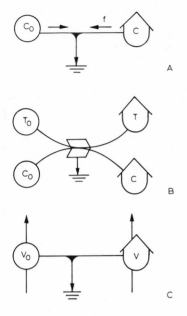

Fig. 6.12. Forms of molecular diffusion. A. Chemical diffusion. B. Thermal coupled chemical diffusion. C. Momentum diffusion.

6.6. ENERGY STORAGE, ENERGY QUALITY UPGRADING, REWARD LOOP PUMPING, AND EDDY DIFFUSION

From Lotka's principle of selection for maximum power comes the corollary that energy gradients with some amount of potential energy beyond a minimum threshold win out in competitive selection if they build potential-energy storages with enough quality to act as amplifiers on the upstream flows serving to pump and accelerate energy capture. Since any energy storage has an inherent rate of entropy increase because of its storage gradient relative to surroundings, there is a drain of potential energy. Since any energy flow generates random noise, it generates continual choice in configurations so that selection for maximum power can proceed rapidly to develop energy reward loops, and the more energy available, the faster the evolution and the stronger the feedback loop system developed. The feedback pumping loops (Fig. 6.4A) result in accelerated energy inflow when the energy gradient is sufficient to generate a storage in excess of its storage loss rate.

Eddy diffusion is a kinetic-energy storage in rotary motion that serves to feed back high-quality, upgraded, potential energy to pump in more, because the eddy acts like a wheel or ball bearing upon which the transport can flow

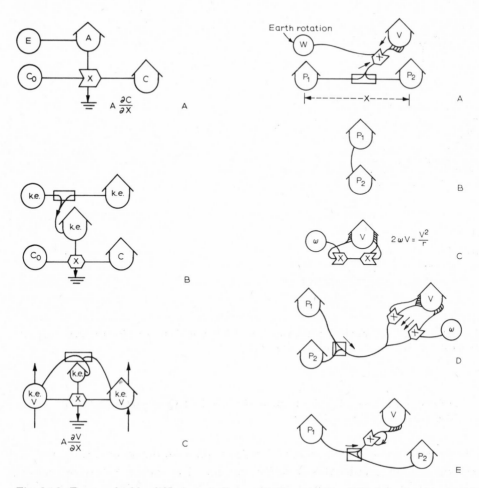

Fig. 6.13. Forms of eddy diffusion. A. Externally forced eddy diffusion of a chemical. B. Chemical diffusion pumped by coupled momentum gradient. C. Momentum diffusion with eddy self-generated. If not shown, velocity vectors are into the paper as in c and d.

Fig. 6.14. Common equilibria in physical oceanography and meteorology. A. Geostrophic balance. B. Hydrostatic balance. C. Inertial motion. D. Meander, gradient wind balance. E. Cyclostrophic motion.

faster than competing pathways. The double-pointed block is the symbol for simultaneous circular facilitation of flow in both directions at once. Recirculating transport carriers are driven by energy stored and maintained in the circular process (Fig. 6.13).

In the consideration of the physical networks of ocean and atmosphere, it is customary to recognize situations in which there are force balances between static forces from energy storage and inertial forces or between inertial backforces. Some common textbook examples are diagrammed in ener-

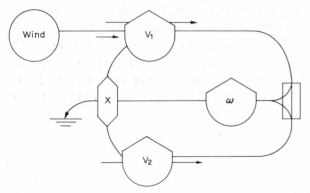

Fig. 6.15. Steady-state balance of wind and frictional force in one dimension (ω denotes eddy motion).

gese in Fig. 6.14. The energy circuit language helps to indicate the sites of energy storage and the force pathways at the same time. Fluid flow occurs along the pathways whenever the forces are imbalanced continuing until velocities change enough to reestablish the force balances. In the real world these systems cannot exist without some additional energy support to match

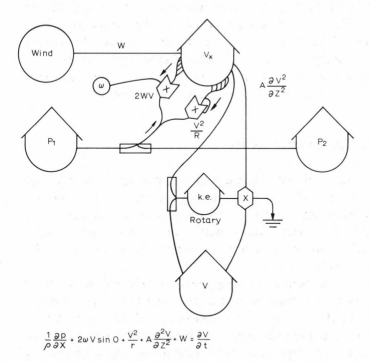

$$\frac{1}{\rho}\frac{\partial p}{\partial X} + 2\omega V \sin 0 + \frac{V^2}{r} + A\frac{\partial^2 V}{\partial Z^2} + W = \frac{\partial V}{\partial t}$$

Fig. 6.16. Energetic circuit diagram of the equation of motion in one dimension with velocity directed into the paper perpendicular to axis of pressure gradient.

that lost into frictional, entropy-increasing, heat dispersion. However, in customary pedagogy, the energy supplements and sinks are not shown in Fig. 6.14. The force balance indicated in Fig. 6.15 is between the component of wind stress and vertical eddy friction opposing the fluid velocity in one dimension without horizontal or vertical terms. The various terms diagrammed in Fig. 6.11—6.15 are combined as a diagram of the equation of motion for one axis (Fig. 6.16).

6.7. FIELDS AND BOXES

The continuous fields of force affecting the fluids of atmosphere and oceans are often visualized in modelling as networks of discrete storages connected by pathways of force and flux, the size of the continuous fluid included in each being an arbitrary decision depending on the detail desired. The sea is thus visualized as a network of repeating duplicate unit models. At each node of the network of similar storages and connectors, there is a set of forces operating, and diagramming one such unit characteristically represents the whole system's operations, each node being similar qualitatively.

At this conference such unit models were referred to as box models. As given in Fig. 6.17A unit model boxes are shown in three-dimensional arrangement. At other times one box is made for the whole system (Fig. 6.17B) without dividing the area into nodes. Simplification in space by overall aggregation leaves some room in the model and computer capacity for complexity other than dissection of fine-grained spatial detail.

6.8. GESTALT IN PHYSICAL OCEANOGRAPHY

In Fig. 6.17C a two-dimensional diagram is given showing a unit model of eddy diffusion of salt (C). If this methodology of repeating connected duplicate submodels were applied to an organism, one would divide the organism into hundreds of grid points with unit models, rather than recognizing the natural divisions in the compartmentalizing of the model into head, heart, digestive system, etc.

Aggregation may have a deeper meaning beyond practical reasons for simplification. The action of natural selection by Lotka's principle of maximum power is not restricted to any size dimension, and the larger dimensions of the sea and atmosphere's motions may have self-organized building structures and motions on a larger scale for useful power maximization. If this is so, then the larger systems of the earth, the oceanic gyres, the cyclones, the general circulation, etc., may be regarded for modelling in the

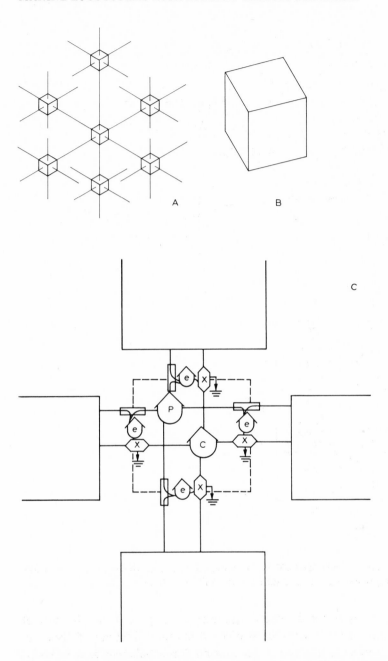

Fig. 6.17. Diagrams illustrating the alternative of modelling a system as one box with overall values for each. A. Model with a network of duplicate unit model boxes connected by pathways of forces, flows, and energies. B. Aggregated model using only one unit model for the overall systems. C. Example of the linking of each unit model to that in the next spacial segment. This example has diffusion of chemical stuffs C under the flows generated by pressure distribution P whose flows generate eddies.

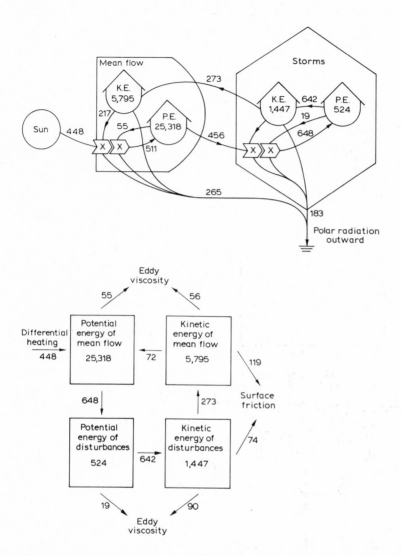

Fig. 6.18. Energy circuit translation (above) of a model of the earth's general circulation given by Hess (1959) (below) quoting Charney and Phillips (1953).

same way that a biologist considers an organism as the proper unit for model recognitions because of unity of organization of energies. Instead of breaking the system into hundreds of nodes calculating force balances and flows in enormous detail with expensive computer manipulations, the system is compartmentalized into only the few main categories of energy storage and flow with emphasis on the causal overall forcing functions, and main feedback mechanisms affecting structure. The trends in recent years to determine overall energy budgets for hurricanes, the earth's general circulation, etc., represent trends toward macro-modelling, although efforts at this kind of modelling in the sea seem few. In Fig. 6.18 an energy budget for the general circulation of the earth given by Hess (1959) from Phillips (in Hess, 1959, p. 347) is redrawn in energy circuit language adding the appropriate symbolism for the forces and feedbacks believed necessary. Note that the empirical study produced the same kind of general energy storages to input pumping as generalized for all systems in Fig. 6.4. This physical system resembles biological ones such as in Fig. 6.1 and 6.2.

Generalizing, one sees the structure of the earth's sea bottoms and shores as built by the energy actions. These forms serve as quality upgraded potential energy, information, and material storages that feed back control actions on the energy input processes. Geomorphological form and information as measured by capital energy investment becomes as much a state variable as biomass in biology or water in hydrology.

In trying to persuade those who work on mechanistic details that there is an overall energy plan in all surviving systems, it is sometimes useful to mention the beach as an example of the potential-energy storage of a system, that builds that physical structure for capturing and channeling further energy toward a stable continuing regime, one adapting to seasonal surges.

In overall macro-modelling and diagramming of physical systems, one questions the usefulness of commonly used catch-all phrase "boundary conditions" referring to various connecting terms in a field of submodels. When one makes energy diagrams, one is forced to distinguish clearly between outside, energy-driven, forcing functions, internal storage state variables, and energy sinks some of which may be energy pumped from outside work actions. Effects of geomorphology become either energy sources if external or storage if internal. The energy diagrams keep the forces and energy considerations under simultaneous consideration rather than as two separate systems.

6.9. POTENTIAL ENERGY BUDGETS FOR PERSPECTIVES; ENERGY QUALITY

Perspective on the importance of components to a system may be gained from energies controlled. Energy diagramming provides a means for converting potential energies in various parts of a system to a common currency often though there are different degrees of energy quality to be recognized. There are two energy values to a flow, one that it carries by itself as its sole energy value as if it is a sole source. The second value is the energy flow that it releases when it reacts at an energy intersection with a second energy source. A high-quality energy has a high ratio of energy released to energy contributed, and this ratio may serve as a quality of energy measure. In

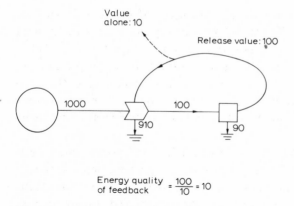

Fig. 6.19. Diagram illustrating energy-quality ratio as the ratio of two energy values that characterize interacting energy pathways.

effect the upgraded high-quality energies are attributed energy values in terms of the amounts of energy of lower quality it can control. For example, in Fig. 6.19 the energy quality of the feedback is 10 kcal (kilocalorie). It is this second value that is useful for impact considerations.

An examination of potential energy in the sea produces some surprises. Given in Table 6.1 is a comparison of the potential-energy inputs to an estuary from physical, biological, chemical, and geological inputs. Solar energy is not potential energy on the same scale as the other high-quality energies until its costs of concentration is included. The fraction of heat from the insolation absorbed, that is potential energy, depends on the Carnot ratio $(\Delta T/T)$ where ΔT is the heat gradient generated. Energy involved in evaporation is mostly reversible energy and not potential energy. Although many people visualize physical energies as large enough to dominate the biological, chemical, and geological aspects this may be a misconception. There is as

TABLE 6.1

Energy inflows to Crystal River Estuary on West Coast of Florida[1]

Energy flow	kcal per m^2 per day
Tidal energy absorbed	0.08
Wave energy absorbed	2.5
Solar energy in photosynthesis	28.0
Solar energy in heating, 2°	27.0[2]
Chemical potential energy in freshwater inflow diluting	66.0

[1] Calculations from Odum, H.T., 1974. Energy cost benefit models for evaluating thermal plumes. Thermal Ecology Symposium Proceedings, Division of Technical Information, U.S. Atomic Energy Commission, Oak Ridge, Tenn. (in press).
[2] Heat flow times Carnot ratio.

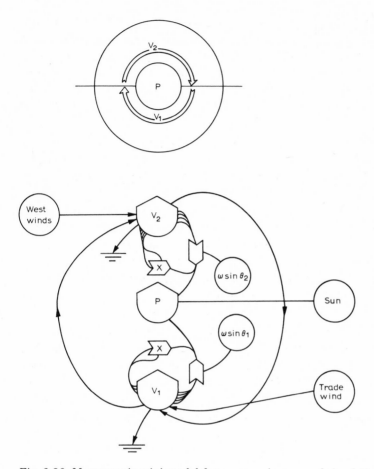

Fig. 6.20. Macroscopic mini-model for an oceanic gyre and simulation of that model.

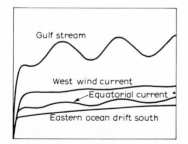

Fig. 6.21. Simulation of model in Fig. 6.20 with sinusoidal inflow program for seasonal winds and heatings. V_2 is West wind current and V_1 Equatorial current.

much potential energy involved in these latter as in the sea's and atmosphere's physical world. The ability of biological and chemical processes to generate controls over physical processes may be greater than often realized. Modelling physical systems without the high storage energy controls of the rest of the system is likely to be erroneous.

6.10. ENERGY COST BENEFIT CALCULATIONS AND IMPACT STATEMENTS

For considerations of value, estimates of high-quality potential energy flow were suggested since the ultimate purpose of any system is later judged by the survivor as good if it was part of survival. Contributions to maximum power and survival thus are measures of value for purposes of maximizing one's economy and ability to compete. Energy diagrams indicate the interactions of factors recognized in proposed environmental changes, and the energies involved served as a measure of relative value. Since the energies involved in many parts of the marine systems are of similar order of magnitude, planning for change in coasts and estuaries that is based on physical properties only may omit consideration of most of the pathways affecting overall value and system survival. It may be incorrect to do piecemeal and subsystem models where they are organized to maximize all the combined energies in coupled and interlaced networks. Approaches to the environmental field made by separate disciplines may show mechanisms but cannot yield plans or predictability if the pattern that survives depends on the whole unified energy maximization.

6.11. A MACROSCOPIC MINI-MODEL SIMULATION*

To illustrate the approach of macroscopic mini-models, a simplified network model for an ocean gyre is given in Fig. 6.20 with forcing functions from sun and wind. In Fig. 6.21 are some simulation graphs from manipulation of forcing functions in seasonal variation. The system generates a rapid current of Gulf Stream type on the left pathway.

REFERENCES

Billig, E. and Plessner, K.W., 1949. The efficiency of the selenium barrier photocell when used as a converter of light into electrical energy. *Philos. Mag.*, 40: 568—572.

---

* Work with H. McKellar.

Charney, J.G. and Phillips, N.A., 1953. Numerical intergration of the quasi-geostrophic equations for barotropic and simple baroclinic flows. *J. Meteorol.*, 10: 71—99.

Hess, S.L. 1959. *Introduction to Theoretical Meteorology.* Holt, Rinehart and Winston, New York, N.Y., 361 pp.

Ivlev, V.S. 1961. *The Experimental Ecology of the Feeding of Fishes.* Yale University Press, New Haven, Conn., 302 pp.

Lumry, R. and Rieske, J.S. 1959. The mechanism of the photochemical activity of related chloroplects. V. Interpretation of the rate parameters. *Plant Physiol.*, 34: 301—305.

Monod, J. 1942. *Recherches sur la croissance de cultures bacteriennes.* Hermann, Paris, 210 pp.

O'Brien, J.J. and Wroblewski, J.S., 1973. A simulation of the mesoscale distribution of the lower marine trophic levels off West Florida. *Preprints of NATO Science Committee Conference, Ofir, Portugal, 1973,* 2: 86 pp.

Odum, H.T., 1971. *Environment, Power, and Society.* John Wiley, New York, N.Y., 336 pp.

Odum, H.T., 1972. An energy circuit language for ecological and social systems: its physical basis. Reprint from: *Systems Analysis and Simulation in Ecology,* 2. Academic Press, New York, N.Y.

Odum, H.T., 1973. Energy, ecology, and economics. *Ambio,* 2 (6) (Royal Swedish Academy of Sciences).

Shinozaki, K. and Kira, T., 1956. Intraspecific competition among higher plants, VII. Logistic theory of the C—D effect. *J. Inst. Polytech. Osaka City Univ.,* DY: 35—72.

Steele, J.H., 1972. Factors controlling marine ecosystems. In: D. Dyrssen and D. Jagner (Editors), *The Changing Chemistry of the Oceans, Nobel Symposium 20.* John Wiley and Sons, New York, N.Y., pp. 209—221.

Verduin, J., 1964. Principles of primary productivity, photosynthesis under completely natural conditions. In: D. Jackson (Editor), *Algae and Man.* Plenum, New York, N.Y., pp. 221—238.

PART II

THE PRESENT STATE OF MARINE MODELLING

CHAPTER 7

PHYSICAL MODELLING

R.W. Stewart

The art and science of numerical modelling of the physics of marine systems is so complicated that most of the literature tends to be too abstruse to be readily accessible to biologists and chemists who may wish to take advantage of it. For this reason it was considered useful that this volume should contain a rather straightforward and simple account of the situation in physical modelling. This article is an attempt to meet that need.

The approach of the physicist to modelling in the marine environment tends to differ significantly from that of his chemical or biological colleague for two important reasons. The first of these is that he believes, and on the whole he is probably right, that he can model his part of the system without the help of the chemist or the biologist. Of course he needs the quantity of dissolved solids in the water in order to be able to calculate the density, but he is so accustomed to making this measurement for himself that he has come to think of it as a physical parameter rather than as a chemical one. Apart from that, he usually needs only parameters which he considers to be within his own domain: temperature, the physical shape of the basin he is working in, driving forces such as tide and wind and perhaps such things as incoming and outgoing radiation, evaporation, rainfall and heat gain or loss to the atmosphere. In a few very special cases he may need some extra information such as the silt load carried by the water and transparency to incoming radiation, but as yet very, very few models are sufficiently refined to be able to make use of this kind of information.

The other factor which influences his attitude is the fact that he is aware that there is an essentially exact set of equations governing the behaviour of the fluid with which he is working. These equations consist of: (1) a dynamic equation (called the Navier-Stokes equation) which describes the momentum balance of his fluid and relates accelerations of the fluid to pressure gradients, the density, gravity, and the existing velocity field; and (2) a continuity equation which in its exact form expresses the idea of conservation of mass, but in oceanography is usually approximated so as to imply conservation of volume. The approximation involved here has been examined carefully and the conditions of its validity (it is nearly always valid in the ocean) are known. He also has some relationship linking the density of the fluid to the quantity of dissolved solids and to the temperature, and he can relate temperature changes to the flow of heat into and out of the fluid.

For most problems this set of equations is in principle closed; that is, given sufficiently well-defined boundary conditions* the problem can *in principle* be solved without injecting any additional hypotheses or assumptions. However, there is a very big difference between saying that the problem can be solved in principle and actually solving it in practice. In practice in most real cases the equations are hopelessly difficult to solve exactly, and additional assumptions or hypotheses must be made use of. Nevertheless, in the few cases where exact solutions can be obtained, and where these solutions prove to be stable, observations agree with the solutions to very great precision. The close agreement which occurs in such cases — so close that there is no discernable difference — colours the attitude of the physicist and leads him to believe that he is dealing with a situation inherently much more amenable to exact mathematical modelling than is biology. (The chemist has his own equally exact relationships, and most physicists are willing to accept chemistry into the field of the so-called "exact" sciences. Few physicists, in their arrogance, would accord such status to biology of any kind, let alone population dynamics!)

In most real cases, however, this arrogant self-confidence of the physicist is ill justified. Among the equations with which he works, some — in particular the Navier-Stokes equation — are non-linear. Non-linear equations have some very unpleasant properties, not the least of which is the fact that in some circumstances small causes cannot be assumed to give rise to small effects. Indeed, there are circumstances in which it can be clearly demonstrated that small causes can ultimately produce effects of comparable im-

---

* Most of the equations of physics are differential equations. By that it is meant that the equations describe relationships between the *rates of change* of some quantities in terms of other quantities. When these equations are solved, the quantity initially obtained is this rate of change. For example, a solution to a particular problem might determine the temperature gradient — that is the rate of change of temperature in space. In order to know the temperature everywhere it is then required only that the temperature at one place be known. Knowing one particular temperature and the temperature gradients everywhere, one can then calculate the temperature everywhere. The specified temperature is called a "boundary condition". The origin of the term comes from the fact that in almost all cases the particular location at which values are prescribed as known is at the edge of the domain — in space or in time — for which the solution is valid. However, there are no rules about this and boundary conditions can quite well be specified in the middle of the domain.

It should be noted that the value of this boundary condition cannot be obtained from the differential equation and has to be inserted on the basis of other information. This requirement for other information is very fundamental to the full solution of differential equations by the use of boundary conditions.

portance to anything else in the system. The most obvious manifestation of the non-linearity is the existence of turbulence. Turbulence arises from fundamental instabilities in the system which exist because of these non-linearities. There is many a situation for which a perfectly good solution to all the equations exists, but flows corresponding to this solution are not usually observed in the real world; the solution corresponds to flows which are fundamentally unstable.

The above words can probably be made more understandable by using a simple example: When an essentially incompressible fluid like water flows through a sufficiently long pipe of sufficiently small diameter and at sufficiently low speed, the profile of velocity within the pipe is very accurately described by a solution to the Navier-Stokes equation, defining what is called Poiseuille flow. This solution to the equations remains valid for all sizes of pipe and at all speeds of flow, but when the combination of these two factors becomes large enough the flow becomes unstable and turbulence is generated. Poiseuille flow, although still a solution to the equations, is not then observed.

The detailed motion of the turbulent flow in the pipe is in a very fundamental way unpredictable. If the boundary conditions were known *exactly*, then in principle one could determine the details of this turbulent flow, since the laws of fluid mechanics remain valid in the turbulent flow. But the detailed exact boundary conditions are in principle unknowable. Not just unknown but unknowable! There are limits to the precision of any measurement and unknowable fluctuations occur within these limits. Because in this system small causes will ultimately give rise to large effects, these originally minuscule fluctuations will eventually become important.

This example is worth pressing a bit further because there is more to be learned from it of relevance to the present discussion. It was stated above that the detailed motion in a turbulent pipe flow is in a very fundamental way unknowable. But that is quite a different thing from saying that nothing can be known about the pipe flow, or nothing predicted with respect to the flow. In fact a very great deal is known and some things can be predicted with great confidence. However, the quantities which are known well are *statistical* quantities, and most of what we know about them has been obtained by empirical measurements in turbulent pipe flows. For example, if we know the roughness of the pipe — defined in a suitable way — then we can predict the *average* velocity at any point in the pipe very well indeed. We can also predict a large number of other statistical properties such as the intensity of each component of the turbulence, the spectrum of the turbulent fluctuations, the rate at which heat will diffuse. We also have information on some more subtle statistical quantities — for example, of the correlation between the instantaneous velocity at one point and that at another, and the probability that a particular speed will not be exceeded over a particular length of time. These statistical properties are exceedingly stable, and where the external parameters — in this case the nature of the pipe and the rate of flow through the pipe — remain fixed, they are very reliable. It should be noted that in detail the pipe flow is full of instabilities, which cause energy flows from one kind of motion to another. It is the averages which are stable. It is worth repeating that this knowledge has been gained almost entirely empirically. There *is* some theory of turbulent pipe flows, and it is getting better. But it remains incapable of yielding predictions about the flow which can be confirmed by observations as well as can the predictions of the Navier-Stokes equation for non-turbulent flows. For very good results one must still rely on a large degree of empiricism. There are some other simple flows, such as circular jets, boundary layers on flat surfaces, wakes behind regularly shaped bodies, which have been intensely studied and for which a comparable body of empirical data exists. From study of the characteristics of these particular flows, plus some examination of more complicated ones, a certain amount of understanding of turbulence has been achieved. But as the complexity of the situation increases, our ability to make confident statements even about statistics decreases.

These facts are important to us, since all of the flows considered in this volume have some turbulent characteristics. Accounting for the effects of turbulence remains the most ubiquitous, persistent, and difficult problem in almost every fluid dynamical situation.

Let us examine a few particular kinds of problems which are treated by numerical models of physical systems. The simplest of these are probably uniform flows in straight channels such as, for example, canalized rivers. In fact, unless there were some complications it would not be worth modelling this kind of situation at all; but complications can occur. For example, there can be significant changes in the rate of flow and one might ask at what flow rate is the system likely to overflow, and also at what rate would a flood crest proceed down the river. Problems of this kind are handled by so-called one-dimensional models. One pretends that the river is divided, in the downstream direction, into a series of boxes*. Each box is the full width and full

---

* This discussion, and all others in this chapter, are couched in terms of the so-called Eulerian description of a system. In an Eulerian approach one chooses an initial grid of points fixed in space and watches changes in the parameters of the flow which occur at these points. The overwhelming majority of existing calculations may be made from an Eulerian viewpoint. There is, however, another way of looking at problems — the so-called Lagrangian viewpoint. In Lagrangian calculations one fixes ones attention on particular parcels of fluid and follows them around as they move in space. Certain kinds of problems, in particular certain kinds of diffusion problems, are more simply formulated in Lagrangian terms, and it is possible that this type of analysis will become more common in the future.

depth of the river. The fluid in a particular box has forces exerted upon it by the difference in elevation of the fluid in the neighbouring boxes and by friction on the sides and bottom of the channel. From these forces one can calculate an acceleration and using a previously known velocity as a boundary condition can then compute a new velocity. The velocity information can then be used to calculate the extent to which a particular box is filling up or emptying, and information is then obtained on the rate of change of elevation in the box. Again, using the boundary condition of a previously known elevation, a new elevation can be calculated. These new elevations are then used again in calculating new accelerations and so forth. The boxes used for calculations of velocities are most usually different from the boxes used for calculating heights, but that is a complication irrelevant to the present discussion. This sort of calculation has been carried out many times with many different systems and has been verified by enough observations that physical modellers have gained a lot of confidence in it. Nevertheless, it should be examined critically with respect to the exact equations described above, which are the principal basis for the physicist's confidence in his methods. Compared with those exact equations, which are differential equations, the numerical equations employed hide a great many things. Instead of having continuous variations in time and in space, the variations come in steps. By making the steps small enough one can overcome difficulties arising from this, but making the steps very small greatly increases the expense of the calculations and there is usually some compromise.

The frictional force is a necessary component for the calculation of accelerations. In systems of this kind the magnitude of the friction is essentially determined by turbulence. We have pointed out that there are no exact methods of calculating the effects of the turbulence, only semi-empirical ones which become less and less reliable as the situation becomes more complex. For the kind of canalized river we are discussing one can do quite well. There is a substantial body of observations on similar rivers and channels if one is working with a hypothetical case (for example, calculating what will happen when a river becomes canalized or calculating what will happen in an artificial channel). In a real case, where one is doing a calculation with respect to an existing canalized river, one can make use of observations. In either situation, frictional force is calculated on the basis of semi-empirical formulae having some support in theory. All effects produced by such things as irregularities in the shape, roughness of the walls, boulders or old bed springs on the bottom, are incorporated in simple semi-empirical expressions. Evidently they cannot be incorporated *exactly*.

There are a lot of other things going on in this canalized river which the model of this kind does not take account of at all. The flow close to the wall and the bottom is much slower than that near the surface or in the middle of the channel, so a single number for the velocity at a point is inadequate to describe the whole situation. These differences are often very important. For example, suppose one is concerned with the transport of a pollutant or of fish larvae down this river. If the source is somewhat at the edge, it will take some time before the substance is diffused right across the river. This diffusion is also a turbulent process and all the discussion about turbulence which we indulged in above is applicable.

Even after the substance has been diffused across the river, the differences in flow rate remain important. Some portions of the substance are transported downstream much more rapidly than others so that in any cross-section — corresponding to a particular box in the model — we find fractions of the substance with wide ranges of age since it left the source. This can be very important in for example the remaining biological oxygen demand of a pollutant or in the stage of development of fish larvae. There will tend to be some non-uniformity in the distribution of properties, in that things found in the middle of the river are likely to be younger than things found around the edges — but this relationship will not be an absolute one since some of the things found in the middle will have spent most of their time travelling near the edges and will have only recently diffused towards the middle, and vice versa.

These complications can also be handled moderately well since again there is a sizeable stock of empirical data. The time taken for material to diffuse across the river can be accounted for by using semi-empirical expressions. The difference in velocity between various parts of the flow can be accounted for by putting in a sort of pseudo-longitudinal diffusion which spreads things out downstream. Again reasonably reliable semi-empirical expressions exist.

If our canalized river is influenced by tide, there is increased complexity because of the back and forth motion, and the fact that the flow does not change directions at all parts of the cross-section simultaneously (the slower parts change direction sooner than the faster ones). The situation is more complex than with the river, depends upon more parameters, and the empirical expressions which have to be used for diffusion and for friction are less reliable. Nevertheless, they remain adequate for many purposes. Modelling of this kind is carried out routinely and successfully.

Appreciably more difficult is similar one-dimensional modelling in an uncanalized river, estuary, channel or embayment. Here one has to deal frequently with mud flats and sand bars which become very shallow or even dry when the water level becomes low. The walls are often indented by coves or have back eddies behind rocks or obstructions. One-dimensional models are unable to account for these things in detail and in some way must parameterize them — an undertaking which is not particularly easy. In hypothetical cases, even the height and flow rate will be calculated with much less precision than for the canalized river. In existing cases, enough observational data can often be taken so that the model can be adjusted quite well. After all, the measurement of the quantity of flow and in particular of the elevation of the surface are fairly straight forward and the results are usually unambiguous. For diffusion studies and calculations of the distribution of substances the problem is tougher. Taking account of the difference in flow velocity across the stream when there are many back eddies and similar phenomena becomes much harder and the empirical injection of diffusion characteristics in such cases is much less reliable than for the canalized river. This data is also much harder to get empirically since the diffusion data is very fundamentally turbulence data and will only have statistical validity.

It is worth labouring this point a bit. Empirical data are needed both to verify our model with respect to flow rates and elevation and with respect to its diffusion characteristics. First consider the question of the elevation: The level across the river varies rather little and one or at most a few measurements of the simple quantity of height above the bottom will establish this level. The empirical data resulting is essentially a single number corresponding to each situation.

Now consider the diffusion situation: The diffusive character of the river varies very widely depending upon where one is in the cross-section of the river. Further, diffusion is a turbulent quantity which can be obtained even in one region only after a large number of measurements, sufficient to establish the statistics of the parameter being measured. Clearly this is a much more difficult, time-consuming, and inherently less accurate procedure than is the measure of water level. Accordingly the semi-empirical expressions used for the diffusion are relatively unreliable compared with those required to determine the height. Nevertheless, useful working models of this kind are fairly common. Given good verification data they can work well in predicting flow volumes and in predicting elevation, and they can give useful if not equally reliable information on the distribution of properties within the system.

Now let us examine the next stage of difficulty. Suppose we have to consider a flow in a body of water which can by no means be considered to be one-dimensional. For example, consider the North Sea, the Irish Sea or one of the Great Lakes.

To model systems like this one must employ two horizontal directions. The usual method is once more to break the system up into boxes, but now the boxes are distributed over the horizontal area. There are some kinds of problems for which these boxes may be taken to reach from surface to bottom, just as in the one-dimensional cases discussed above they were taken to encompass the whole cross-sectional area of the river or channel. If one is dealing with a relatively rapidly moving wave, with wave length much longer than the water depth, this approach can be used successfully. Examples include studies of tides and of storm surges. There are some tidal studies which include the whole world ocean in this fashion. In that case the driving forces are the astronomical ones. Others are more restricted, for bodies such as the North Sea, the Irish Sea or the Strait of Georgia. A tidal model for bodies of this kind must be driven by influences imposed at the openings of the body, for example in the North Sea at the northern boundary and at the Straits of Dover. In such cases the boundary conditions are of overwhelming importance, and more clearly than is usual it can be seen that they determine what goes on in the system.

Turbulent boundary friction is usually important and must be introduced semi-empirically.

Many tidal models of this sort have been constructed with considerable success. However, it must be recognized that a large factor in the success is the rather readily obtainable empirical data on tidal heights, and to a lesser extent on tidal currents, which can be used to improve the accuracy of semi-empirical parameters employed.

Additional driving forces, such as winds and rivers flowing in through the boundaries can be incorporated in such models. They can be used for example to study wind-induced seiches and storm surges in bodies of water. The flow field determined by these models can be used to give information about the distribution of pollutants and other components in somewhat the same way as was discussed above for the case of the channel. Once more, additional semi-empirical information has to be put in about lateral diffusion, and this information is on the whole even less secure than that for the one-dimensional case so the results are cruder. Where winds or rivers are present, there result additional differences in velocity between the surface and the deeper water, which further complicate the problems of trying to describe the distribution of substances in the water.

Another kind of two-dimensional model may be employed in narrow channels and deep estuaries, where the two dimensions are "downstream" and vertical. For many purposes, fjord-like estuaries demand such treatment. In a fjord one typically has heavy salty water of marine origin underlying brackish or much diluted water, strongly influenced by incoming rivers, at the surface. Few characteristics of a system of this kind can be successfully described by a one-dimensional model, since the difference between the deep water and the shallow water is absolutely crucial to its behaviour. So is the mixing between these two water masses. Typically the energy for this mixing is largely supplied by the tide, which must therefore be incorporated into the model in some way. However, the effectiveness of the mixing is determined crucially by the degree of density stratification, and once more semi-empirical information has to be added. Many phenomena of systems of this kind cannot be understood without consideration of the wind. Wind effects can dramatically change the depth of the upper layer in quite a short time.

Situations of this kind frequently demand consideration of internal waves and internal seiches, which can exist when the water column is stably stratified vertically. Internal waves are fairly well understood and existing techniques can deal with many of their characteristics quite successfully. However, they are dissipated by turbulent-friction mechanisms and by wave-breaking mechanisms neither of which are particularly well understood, and in being dissipated they generate turbulence which is effective in mixing other properties in the water. Once more semi-empiricism has to be resorted to in incorporating such effects into models. Such efforts are in their infancy.

Having considered systems in two dimensions horizontally, and two-dimensional systems in which one dimension is vertical, the logical development, of course, is to consider the full three-dimensional model in all its glory. Three-dimensional models are employed with some success in meteorology. In fact they are the basis for modern numerical weather forecast methods. A few three-dimensional models have been constructed for marine situations, frequently by adapting existing meteorological models. The three-dimensional model would seem to be the ultimate objective of all modellers. But typically it demands a lot of time on a very large computer so, reasonably, most workers attempt to make do with something less whenever possible. For example, a favourite approach is to consider that the body of water is a two-layer one with lighter water — lighter because it is warmer or less saline or both — separated from the deep heavier water by an artificially sharp interface. Some problems may be examined this way. Others, for example, anything involving the vertical propagation of internal waves, or the intrusive flow of fluid of intermediate density, cannot be so handled. One of the serious holdups preventing the more rapid development of three-dimensional models of marine systems is the lack of the requisite data base both for initialization and verification of models. As has been repeatedly pointed out in this discussion, the fluid dynamicist must over and over again fall back upon semi-empiricism. Except in very simple situations, he can have only limited confidence in his semi-empirical constants and functions.

An example will again serve to clarify this situation. Suppose the objective of a model is to examine the distribution of pollutants in the North Sea originating from the Rhine River. To accomplish this the model has to account for the mean currents in the sea produced by tide, winds, and effects introduced at the boundary such as by density differences between water in the English Channel and water at the northern boundary, and the influence of rivers flowing into the sea. It must also take account of the diffusion processes. Now in the third paragraph of this article it was pointed out that the fundamental dynamic equation — the Navier-Stokes equation — determines acceleration in terms (among other things) of the existing velocity field. Thus ideally the model should be initiated, i.e. given a boundary condition in time, with a specified velocity corresponding to the real field at some instant. The model would then be allowed to run for some time, after which it will predict a new velocity field. Ideally one would like to check that velocity field against a new observed one, observed after a corresponding time interval. This is precisely what is done in meteorology, where the model is initiated on the basis of the large body of data collected by the World Weather Watch, and checked, i.e. verified, by comparison with data taken at a later time by the same observing system. However, nothing comparable to this observing system exists in any large body of water. Accordingly ability to improve and refine models is significantly less than in the meteorological case, and the confidence one might have in the predictions of these models is correspondingly less.

So far as diffusion is concerned, the situation is not very good in meteorology and is even worse in oceanography. As has been pointed out above the requisite diffusion data is both harder to get and inherently less reliable than

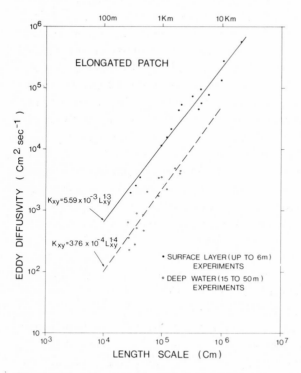

Fig. 7.1. Variation of eddy diffusivity with length scale (from Murthy, 1972).

those on mean flows. Diffusion is usually modelled as though it were analogous to molecular diffusion. That is, the rate of diffusion is assumed to be proportional to the gradient of the quantity diffused multiplied by a factor called the eddy diffusivity. There are good theoretical reasons for believing that this eddy diffusivity should increase in magnitude as the scale of the phenomenon being considered increases. Most observations confirm this. A typical result from a good, carefully performed, observational program is shown in Fig. 7.1 taken from Murthy (1972)*.

There one can see several things. The eddy diffusivity increases markedly with scale, and within a factor of about two follows a semi-empirical law quite well. Note that the diffusion varies by a factor of ten between the near-surface water and water only a few tens of meters deeper. It is evident that an accurate incorporation of diffusion into a three-dimensional model is both difficult and requires a great deal of data. It is not easy theoretically to accommodate the fact that the eddy diffusivity depends on scale. To take account of the fact that its magnitude varies dramatically with the depth calls for a very great deal of empirical data.

All this is to say that we are not helpless; we know a great deal already, but we need to know a great deal more before fully reliable three-dimensional models can be used with confidence by those who need them.

---

* Murthy, C.R., 1972. *Canada Centre for Inland Waters, Annual Report, 1972.* Burlington, Ont.

CHAPTER 8

CHEMICAL MODELLING

Edward D. Goldberg

8.1. INTRODUCTION

Marine chemistry encompasses both a description of the chemical species within the world's oceans and their interactions not only among themselves but also with the components of the atmosphere, sediments and biosphere. An understanding of marine chemical dynamics requires a knowledge of the reactants and their products. The complex nature of the multi-component electrolyte solution sea water has been interpreted in terms of ion-pairing involving short-range interactions between the dissolved species. Since nearly all the stable elements have been found in sea water, the number of possible chemical species is quite high. So far, the most extensive speciation studies have involved a dozen or so of the elements in highest concentrations. Our descriptions of sea water components have been primarily restricted to systems at temperatures and pressures of the laboratory (say 25°C and 1 atm.) even though the ranges of such variables in nature span three orders of magnitude. The overwhelming complexities of the sea water system have been partially overcome in recent years through computer models for the speciation studies.

The aim of this presentation is to provide an overview of modelling in marine chemistry through an interweaving of some recent studies with the papers presented at the NATO Conference. There is no attempt to be exhaustive or even to cover all facets of marine chemical modelling. The brief summary of past work provides a background to consider the Conference contributions.

One of the cliches of marine chemistry is the concept that chemical reactions take place at interfaces: water—sediment, water—air and water—organisms. As a consequence, chemical systems have been defined using these simplistic, but ofttimes useful categorizations. Box model simulations of the oceans which extend from two component systems (a mixed and a deep layer) to twenty or so component systems (involving mixed and deep layers for each of the ocean basins as well as the atmosphere and sediments) have made the study of global chemistries tractable.

As a result of man's mobilization of materials about the surface of the earth, he has sometimes inadvertently and sometimes deliberately altered the chemical make-up of sea waters. Some of these activities have made studies of natural processes more difficult; others have provided chemical tracers whose environmental behaviors have given greater insight into natural processes. For example, the reactions of many trace chemical species at or below micromolar concentrations in sea water have become more fully understood by studies of the artificial radioactive nuclides introduced as a consequence of nuclear device detonations or of the discharge of wastes from nuclear reactors.

8.2. CHEMICAL SPECIATION

Our present understanding of the chemical nature of sea water involves interactions between its dissolved ions and molecules. For example, no longer do we consider sulphate ion as the only form of sulphur in solution in aerated waters, but we conceive of a host of ion-pairs resulting from chemical binding of sulphate ion with such species as calcium ion, magnesium ion, and copper ion. These reactions are described by equations of the type:

$$M + L = ML$$

where M is the cationic species, L a negatively charged or neutral ligand and ML the complex ion or ion-pair. Under equilibrium conditions, the distribution of these three species is given by:

$$K = a_{ML}/a_M a_L$$

where K is the equilibrium or association constant and the a's are the activities of the subscripted species. The activities are the product of the concentration of the species under consideration and the activity coefficient, which is a function of the salinity, temperature and pressure. Two approaches have been made to apply these considerations to the sea water system: (1) determining the association constants in media with the same ionic strengths as sea water — in such a case the activity coefficients are included in the "apparent equilibrium constant"; and (2) using the association constants obtained for infinite dilution with activity coefficients estimated for the sea water system. Both techniques should yield equivalent results.

The first model of such interactions for the major ions of sea water was provided by Garrels and Thompson (1962). For 1-atm. pressure and for 25°C, they considered the species formed by Na^+, K^+ Mg^{2+} and Ca^{2+} with SO_4^{2-}, HCO_3^- and CO_3^{2-}. Only 9% of the carbonate was estimated to be present as the free ion with 7% associated with calcium, 67% with magnesium and 17% with sodium. Similarly, about 40% of the sulfate ion was bound up with magnesium and sodium as ion-pairs with 54% existing as the free ion. Subsequently, there have been made alterations to the original model usually involving what appear to be more reliable equilibrium constants or more reasonable values for the activity coefficients (Kester and Pytkowicz, 1968; Dyrssen and Wedborg, 1974). In addition, Sr^{2+}, H^+, F^- and $B(OH)_3$ have been included among the major species and their interactions considered in more elaborate models.

8.3. THE MAJOR SEDIMENTARY CYCLE

Over the past fifty years several models have been put forth to quantitatively describe the amounts of crustal material weathered and of the dissolved and solid phases transported to the oceans. Such models can be conveniently divided into two categories: integral and differential. The former considers a material balance for the elements over the time period of the existence of the present oceans, whereas the latter seeks out present-day inputs and outputs in the weathering cycle.

The principal integral approach, first elaborated by Goldschmidt (1933) and subsequently described as "the method of geochemical balances", involves a model in which crustal rock is decomposed and the remnants are accommodated in two reservoirs: the dissolved species in sea water and the sediments. A series of material balance equations can be set up:

$$\sum_i x_i^a X_i = \sum_j y_j^a Y_j + z_i^a Z$$

where X_i is a type of crustal rock, Y_i is a type of sediment derived from the weathering of crustal rocks and Z is the total mass of elements in sea water. The coefficients x_i^a, y_i^a and z_i^a refer to the weight fractions or average concentrations of element a in crustal rock types, in sediments, and in the dissolved state in the ocean, respectively.

The solutions of the above equations have been carried out using different combinations and numbers of elements from the original work of Goldschmidt (3 elements) to Horn and Adams (1966) whose computer solutions involved 65 elements (Table 8.1). In spite of the increased sophistication of the models over the years, the range in the amount of weathered igneous materials has only varied between 740 and 2040 billion tons.

Recently, overall models for the cycling of chemical elements through reservoirs at the surface of the earth, based upon their involvement in mineral formation, have been put forth (see for example Garrels and Perry, 1973). Such models, involving the major elements, emphasize the influence

TABLE 8.1

Comparison of geochemical balance calculations as a function of time and numbers of variables (Horn and Adams, 1966)

Elements used in material balance computations	Estimate of weathered igneous materials (10^{15} metric tons)	Reference
Na, Ca, Mg	815	Goldschmidt (1933)
Na, Si	740	Conway (1943)
Al, Fe, Na, K, Si, Ca, Mg	989	Wickman (1954)
Na, K, Ca, Mn	941	Goldberg and Arrhenius (1958)
65 elements	2040	Horn and Adams (1966)

of mineral formation upon the composition of the atmosphere and hydrosphere.

Perhaps a more substantial entry into the book-keeping of the major sedimentary cycle can be made on the basis of what is taking place today — the differential methods. Computations of the rate of land erosion were initiated by Clarke (1924) who used equations of the type:

$$\text{rate of land erosion} = \frac{\text{annual input of river water per year to the oceans}}{\text{drainage area}} \times \frac{\text{solid load of rivers, average}}{\text{average density of crustal rock}}$$

Clarke considered only the chemical denudation of exposed rocks through solution processes. Using the best available data on river compositions and annual discharges, he estimated the lowering of land by the solvent action of water to be of the order of 1 cm/1000 years.

There have been subsequent computations of erosion rates where usually both the dissolved and solid loads of rivers are considered (see for example Judson, 1968). Values somewhat higher than those of Clarke are found; for the regional erosion in the United States 5 cm/1000 years is typical, whereas other parts of the earth range both slightly above and below this value. An interesting consequence of the Judson model is that the rate of erosion is 2.5 times greater since man involved himself in the geological cycles.

Another type model which allows an entry to book-keeping the major sedimentary cycle on a differential basis utilizes the residence time, defined as the average time which a chemical species spends in a component of the environment. In the simplest case, the oceans are considered a single domain and the residence time is given by $T = A/(dA/dt)$ where A is the total amount of the element in suspension or solution, and dA/dt is the amount introduced or leaving per unit time. The concept is based upon an assumption of steady state and of a well-mixed ocean. Elaborate steady-state models involving many components have enlarged our knowledge of large-scale oceanic mixing processes and their impacts upon the distributions of natural and man-generated chemical species (see for example Broecker, 1963).

Fewer attempts have been made to model the non-steady-state situation in which materials are introduced on a regional or global basis to the marine environment. Such cases usually involve pollutants generated by man's activities. Besides the models developed for the PCBs (polychlorinated biphenyls) and DDT residues considered in a following section, there is the model formulated by Tatsumoto and Patterson (1963) to simulate the distribution of lead about the environment resulting from the combustion of tetraethyl leads in gasoline engines. This abnormal rate of lead introduction, based upon use statistics in the Northern Hemisphere, caused an increase in surface ocean waters. Predictions of surface water concentrations based upon a non-steady-state model were in agreement with observed concentrations. Biological removal of lead to depths determined its residence time in the surface waters.

8.4. AIR—SEA INTERACTIONS

Estimates of the material fluxes from the atmosphere to the oceans and from the oceans to the atmosphere have been dependent upon models which reduce highly complex mechanisms, often taking place at the molecular level, to tractable schematizations. The flow of solids to and from the oceans has been reviewed recently by MacIntyre (1974) and Goldberg (1971). In the former survey, emphasis is placed upon chemical fractionation at the interface during the ejection of particulate matter from sea to air. The latter presentation considers the atmospheric transport of naturally mobilized and man-mobilized materials from the continents to the oceans. There is an extensive literature on evaporational phenomena and more recently emphasis has been on gas transfer. Our discussion will consider a recent attempt to computer model the carbon dioxide distribution at the earth's surface, where the oceans play a primary role, and its relation to one of the contributed papers to the Conference, a model of oceanic gas-exchange processes. In addition, we will review several other models involving vapor phase transport of man-generated materials.

A concern of man's alteration of his environment involves the production of carbon dioxide from the combustion of fossil fuels, coal, oil, and natural gas. At the present time the CO_2 content seems to be increasing at a rate of about 0.2% per year or 0.7 p.p.m. out of the 320 p.p.m. of CO_2 in the atmosphere on the volume basis. There exists the possibility that the additional CO_2 may have a long-term effect upon climate, possibly as a result of its adsorption of radiation and a heating of the upper atmosphere. There are three reservoirs for carbon dioxide at the earth's surface: the atmosphere, the biosphere and the oceans. To predict future levels of carbon dioxide in the atmosphere as a result of inputs from fossil fuel burning, several models have been devised, one of the more elaborate ones due to Machta (1972). The fundamental query relates to the distribution of carbon dioxide with time between the three sinks. The ability to foretell future carbon dioxide levels in the atmosphere depends upon a knowledge of: (1) the input of carbon dioxide from the burning processes; (2) the exchanges of CO_2 between each reservoir and within components of each reservoir; and (3) the stability of each reservoir. As in most models of this type, the present provides a key to the future, and the characteristics of the reservoirs are assumed to be unalterable with time. The model used is shown in Fig. 8.1. First-order kinetics are assumed in exchanges of gas between the troposphere and stratosphere, between troposphere and mixed layer of the ocean, and between mixed and deep layers of the ocean.

A novel part of the model involves the definition of the biosphere as a reservoir into which a mass of carbon is transferred each year equal to the net primary production. This transfer can take place from the troposphere to the terrestrial biosphere or from the mixed ocean layer to the marine biosphere. The prediction of the CO_2 in the atmosphere, as a function of time to the year 2000, is shown in Fig. 8.2. The fit of observed and predicted values between 1958 and 1971 is almost perfect. The model also allows an estimation of the partition of carbon dioxide between the various reservoirs. In 1970, for example, 55% of the carbon dioxide produced in fossil fuel

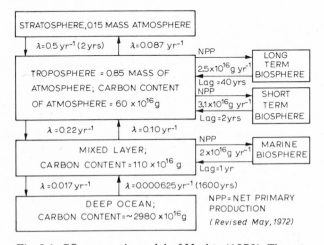

Fig. 8.1. CO_2 reservoir model of Machta (1972). The rate constants are first order.

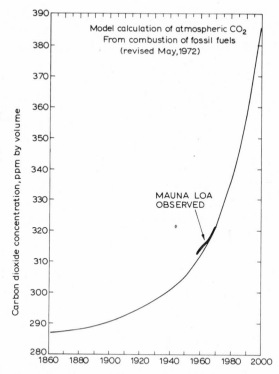

Fig. 8.2. The predicted history of the atmospheric burden of CO_2 (Machta, 1972). The inset compares predicted values with those observed at the Mauna Loa Observatory in Hawaii.

combustion up to this date resided in the atmosphere; 15% in the biosphere and 30% in the oceans — 25% in the mixed layer and 5% in the deep layer.

A second feature of the carbon dioxide distribution is its seasonal variation. In the Northern Hemisphere there is a pronounced decrease during the five summer months with a slow recovery during the remaining seven. A much smaller variation between summer and winter is experienced in the Southern Hemisphere. These results, fitted to the model, can be explained by seasonal changes in the biospheric uptake and release of carbon dioxide, primarily on the continents.

In another vein involving man's impact upon the ocean system via the atmosphere is the unexpected behavior of the halogenated hydrocarbons, DDT and its metabolites, and the PCBs. Both of these collectives of compounds have wide distributions in the ocean system, following their use by man, an observation that was difficult to reconcile with their unusually low vapor pressures ($1.5 \cdot 10^{-7}$ mm of Hg for DDT and 10^{-4} to 10^{-6} mm of Hg for the PCBs). However, their rather high concentrations in atmospheric dusts (Risebrough et al., 1968) and in rains (Bevenue et al., 1972) directed

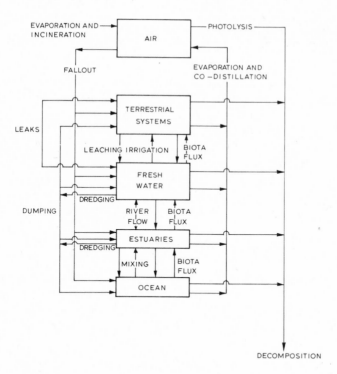

Fig. 8.3. Environmental transport model of Nisbet and Sarofim (1972) for PCBs.

many investigations to the entry of these substances to the oceans via the atmosphere (Harvey et al., 1972, 1973).

Models to explain the widespread distributions of PCBs (Nisbet and Sarofim, 1972) and DDT and its metabolites (Woodwell et al., 1971) have been presented. Nisbet and Sarofim examined the possible routes of PCBs into the environment on the basis of both use and disposal information, coupled to production figures (Fig. 8.3). Their environmental transport model predicted 15 kilotons of PCBs had accumulated in the North Atlantic, in part by localized discharge and in part by aerial fallout, in the early 1970s. Harvey et al. (1973) on the basis of actual measurements in these waters estimated them to contain 18 kilotons of PCBs in 1972.

A micro-scale model for the gas exchange between air and sea water by Liss and Slater (1973) has some most consequential applications on a global basis. A two-layer film system is put forth as the interface between the air and water, both of which are assumed to be well mixed. The principal resistance to gas transport arises from the interfacial layers, where gas movement takes place by molecular processes (Fig. 8.4). Using Fick's First Law of Molecular Diffusion and assuming that the transport of the gas across the

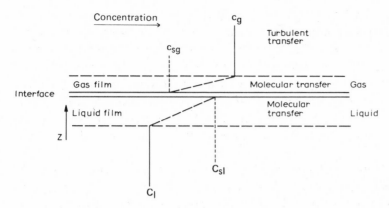

Fig. 8.4. Two-layer model of Liss and Slater (1973) for the air—sea interface.

interface is a steady-state process:

$$F = k_g(c_g - c_{sg}) = k_l(C_{sl} - C_l) \tag{8.1}$$

where k_g and k_l are the diffusion constants for the gas and liquid phases respectively and the concentrations refer to the subscripted phases. If the exchanging gas obeys Henry's Law, then:

$$c_{sg} = HC_{sl} \tag{8.2}$$

and eqs. 8.1 and 8.2 can be combined into the following with the elimination of C_{sl} and c_{sg}:

$$F = \frac{c_g - HC_l}{1/k_g + H/k_l} = \frac{c_g/H - C_l}{1/k_l - 1/Hk_g} \tag{8.3}$$

This equation can be rewritten as:

$$F = K_g(c_g - HC_l) - K_l(c_g/H - C_l) \tag{8.4}$$

where

$$1/K_g = 1/k_g + H/k_l \tag{8.5}$$

and:

$$1/K_l = 1/k_l + 1/Hk_g \tag{8.6}$$

The total resistance (R_t) to exchange, for example on the basis of the liquid phase, is given by:

$$(R_t)_l = r_l + r_g \tag{8.7}$$

with $r_l = 1/k_l$ and $r_g = 1/Hk_g$. Individual resistances are most difficult to measure in field and laboratory studies; so far only the total resistances have been determined. Further, Liss and Slater argue that the liquid phase resistance is negligible for water molecules crossing the air—sea interface and that the exchange will be controlled by gas phase processes. Liss (1973) determined k_g for water as about 300 cm/h and obtained the value for other gases by multiplying this figure by the ratio of the square roots of the molecular weights of H_2O and the other gas. The gas phase resistances were gathered from the literature involving investigations with sea water. A mean value of k_l = 20 cm/h appeared reasonable for gases with molecular weights between 15 and 65.

Several other factors influence the liquid phase resistance. Where the gas is involved in chemical reactions with solute species, the gradients of the products of such reactions in the liquid phase can affect significantly the exchange reactions. Thus, the pH, which determines the $(HCO_3^-)/(CO_3^{2-})$ ratios, influences the exchange of CO_2 across the surface. In addition, the CO_2 exchange is related to the thickness of the interfacial liquid layer and the hydration rate constant. Under conditions of turbulence the thickness of the liquid film decreases. Thus, the observation that k_l increases approximately as the square of the wind velocity is not an unexpected finding. These observations may have significant implications to the global carbon dioxide budget. Changes in weather, such as the intensity of prevailing winds as a result of changes of atmospheric CO_2 levels, in principle can affect the ocean's ability to act as a sink. The inclusion of the gas exchange equations developed by Liss and Slater and their predecessors in this area clearly can improve our predictability of the future CO_2 levels in the atmosphere.

A second use of the models is the estimation of gas fluxes across the air—sea interface. Appropriate values of H along with k_g = 3000 cm/h and k_l = 20 cm/h were introduced into eqs. 8.5 and 8.6. The data are summarized in Table 8.2. These figures are in reasonably good agreement with flux estimations arrived at in other ways. The sulfur dioxide flux is of the same order of magnitude as the estimated inputs of either anthropogenic (fossil fuel combustion) or natural processes. For the group of gases, N_2O, CO, CH_4, CH_3I, and $(CH_3)_2S$, the ocean is computed to be a source, in line with recent investigations. On the other hand, carbon tetrachloride and the exten-

TABLE 8.2

Gas fluxed to and from surface ocean water

Gas	SO_2	N_2O	CO	CH_4	CCl_4	CCl_3F	MeI	$(Me)_2S$
Total oceanic flux g/year	1.5 $\times 10^{14}$	1.2 $\times 10^{14}$	4.3 $\times 10^{13}$	3.2 $\times 10^{12}$	1.4 $\times 10^{10}$	5.4 $\times 10^9$	2.7 $\times 10^{11}$	7.2 $\times 10^{12}$
(+ sea → air) (− air → sea)	−	+	+	+	−	−	+	+

sively used low-molecular weight fluorocarbons appear to be entering the oceans from the atmosphere. The flux calculations are in remarkably good agreement for all of these materials with field studies. Clearly, the model has great predictive value for gases produced and mobilized by man. Not only can atmospheric values be predicted, if estimates of production loss to the environment are available, but also those for surface ocean waters, where undesirable effects upon life process might occur.

8.5. SEDIMENT—WATER INTERACTIONS

The deep-sea floor is usually conceived of as a sink for materials precipitating out of sea waters, yet there are instances in which it is a source. For example, the high concentrations of radium in deep-sea waters, relative to surface values (Koczy, 1958) have been attributed to a diffusion of the element from the sediments where it is produced in the natural uranium and thorium decay series. Models to study the migration of chemical species have been proposed with varying degrees of sophistication.

One of the first (Goldberg and Koide, 1963) models had the following formulations. The sediments were assumed to have a uniform rate of accretion and contained invariant concentrations of ^{230}Th, the parent nuclide of the diffusing radium isotope ^{226}Ra. On such a basis the concentration of ^{226}Ra at any depth Z in the sedimentary column is given by:

$$n(Z) = \left[\frac{\lambda_1 N_0}{K(\lambda_1/C)^2 + \lambda_1 - \lambda_2} \right] \left[\exp \frac{(C - \sqrt{C^2 + 4K\lambda_2})Z}{2K} - \exp\left(-\frac{\lambda_1}{C} Z\right) \right.$$

$$\left. + n(0) \exp \frac{(C - \sqrt{C^2 + 4K\lambda_2})Z}{2K} \right]$$

where λ_1 is the decay constant of ^{230}Th, λ_2 is the decay constant of ^{226}Ra, K is the migration coefficient of ^{226}Ra, N_0 is the concentration of ^{230}Th in freshly deposited sediments, C is the velocity of accumulation of sediment, and $n(0)$ is the concentration of ^{226}Ra in freshly deposited sediment. The values of K determined experimentally were of the order of 10^{-9} cm^2/sec, four orders of magnitude less than the diffusion constant in water. These low values of K suggest that simple diffusion does not explain the migration of radium from the sediments to sea water. It appears that sorption reaction of radium on the surfaces of the sedimentary solids limits the migration of radium.

Subsequently, more sophisticated models of diffusion in sediments have appeared. Many of these included compaction effects but did not consider any advective terms. Tzur (1971) developed a formulation which included both diffusion and advection of solutes in accumulating sediments. He takes into account the compaction of the sediments and the advection induced by it. His general conclusion was that the advective terms could in general be eliminated, but cases could arise where they would be important. For instance, where sediment accumulation rates are greater than 10^{-11} cm/year, convection may be more important than diffusion and can prevent the diffusion of solutes from sedimentary domain to the interface. With the development of in situ interstitial water samplers (Sayles et al., 1973; Barnes, 1973) these models may now be evaluated with field data.

8.6. BIOTA—WATER INTERACTIONS

The most dramatic differences in the chemical compositions of marine waters, either spatially or temporally, result from the primary productivity of organic matter in surface waters and the subsequent combustion of this material at greater depths. There have been many efforts to model the primary productivity process in the oceans where its intensity is related both to physical and chemical parameters. Fewer efforts have been made in simulating the distribution of chemical elements in various spheres of the marine system where the biological processes play a significant but clearly not the only influential role. The development of marine chemistry during the second quarter of this century was centered primarily about the nutrients (phosphorus, silicon, nitrogen and vitamins) required in plant growth. Their assay in marine waters was an accepted part of most oceanographic expeditions. An enormous volume of data, good and bad, of nutrient concentrations in the world ocean evolved. More is being accumulated. At the present time the first attempts to collate nutrient concentrations for the world ocean or for major parts of it into a coherent picture through computer modelling are being made.

But what of the other elements whose oceanic concentrations are influenced by marine organisms? Many metals are known to be highly enriched in both plants and animals of the sea. Downward transport and subsequent decomposition of the particulate remains of marine organisms have been invoked to explain the higher concentrations of such elements as barium, the rare earths, and thorium in deep as compared to surface waters. However, at the present time there is a paucity of data for the marine concentrations of these heavy metals, often at sub-micromolar concentrations. Hence, models to simulate their distribution about the marine environment are as yet impossible to formulate.

An area of controversy that may be resolved by appropriate modelling concerns the relationships, if any, between the distribution of particulate and dissolved organic matter in deep waters to the biological processes in surface waters. There is a growing base of data being accumulated with apparently reliable techniques. Menzel (1974) suggests that our inability to interpret distributions arises from the lack of rates for the cycles of organic matter, except that involved in primary productivity. Modelling may allow a delimiting of possible values for these unknown rates.

Phosphorus is perhaps the most studied of the inorganic nutrients involved in marine primary productivity. Adam (1973) and Postma (1971) have each contributed a paper concerned with the cycles of phosphorus in marine systems. Adam constructs a model for a shallow sea where interactions with the sediments are significant, while Postma dedicates himself to an understanding of the transport processes and reservoirs of phosphate in an ocean basin.

Adam considers five forms of phosphate: (1) dissolved phosphate species; (2) non-living particulates containing phosphorus; (3) sedimentary phases containing phosphorus; (4) dissolved phosphate species in interstitial waters; and (5) the phosphorus in the biosphere — there are two subsets here: the plankton and the benthos. Three types of interactions are considered:

(a) *Biological interactions* involving the dissolved phosphate species. Death of these species releases phosphorus in the particulate form. The benthos uses the dissolved phosphate in the interstitial waters of the sediments.

(b) *Physico-chemical interactions* include dissolution of particulates, formation of particulates, sedimentation of particulates suspended in the water column, a subsequent resuspension and diffusion of phosphate species through the sediment—water interface.

(c) *Bacterial activity* in which there occurs the dissolution of the particulates containing phosphorus and their subsequent release.

In addition, it is assumed that there is no predation of the benthos upon the plankton or of the plankton upon the benthos.

The equations take the usual forms. For the dissolved phosphate species, subscripted 1, the change in amount with time, x_1 is given by:

$$\dot{x}_1 = I_{12} - I_{21} + I_{13} - I_{51} + I_{15}$$

where I_{21} is the formation of particulate phosphorus in sea water from dissolved forms; I_{12} is the dissolution of particulates containing phosphorus; I_{51} is the uptake of dissolved phosphate species by the benthos and plankton; I_{15} is the release of dissolved phosphate by the benthos and plankton; and I_{13} is the diffusion of dissolved phosphate from the interstitial waters. Other interactions are assumed to be unimportant. The final term is the only one that may be negative, although Adam points out in their searches it is always positive, i.e., the interstitial waters provide phosphate species to the oceans.

Adam provides significant novelty to his model through the use of non-linear interaction terms. For example, his biological uptake term I_{51}, which is given by $I_{51} - R_{51}x_5$ where x_5 is the mass of phosphorus in the biosphere and R_{51} is the interaction term, is a non-linear function. Adam argues that there are two types of phosphorus in the primary producing plankton: labile and fixed varieties. Excess labile phosphate is presumed to reduce the rate of growth. The interaction term is presented as:

$$R_{51} = F_{51} \frac{1}{\rho_5 + c_5}$$

where ρ_5 is half of the value of phosphate in the biosphere at saturation, c_5 is the mass of phosphorus in the biosphere divided by the total mass of water to which it is exposed and:

$$F_{51} = \alpha_{51} \frac{c_1}{c_1 + \sigma_1}$$

with c_1 being the phosphate concentration in the water, and α_{51} and σ_1 being biological and physico-chemical constants, respectively.

Many of the coefficients, such as F_{51}, have not been well determined from field or laboratory investigations and their orders of magnitude have been approximated from the sparse data in the literature. Some were purely guesses. Nine case studies were carried out as a function of differing concentrations in the reservoirs and of differing values to the biological and physico-chemical constants. One of the cases, in which the interaction terms are nearly linear with respect to the physical variables, is given in Chapter 5 (Fig. 5.5). In this example, the amount of phosphorus in the plankton and benthos (Reservoirs 5 and 6) first show a rapid uptake of phosphorus. With a decrease of the phosphorus in the aqueous reservoir, the rate of uptake in the biosphere decreases. Further bacterial decomposition of organic detritus enters the picture with increasing time. There is a loss of phosphorus from the sediments due to diffusion into the overlying water column. The particulate phosphate increases with time due to its production from the growing biomass. With time, its rate of increase decreases. The dissolved phosphorus in the water, after going through a minimum with plant growth, reaches a plateau reflecting a near steady-state condition. The power of this approach rests with the availability of field data which will allow an enumeration of the most reasonable values for the biological and physico-chemical constants.

Postma develops the distribution of phosphate in the world ocean with a consideration of the following processes: circulation, biological uptake and mineralization, the conveyance of phosphate to greater depths with organic debris followed by a regeneration to the water as a consequence of oxidation, the migration of animals and supply from land. In the open ocean the

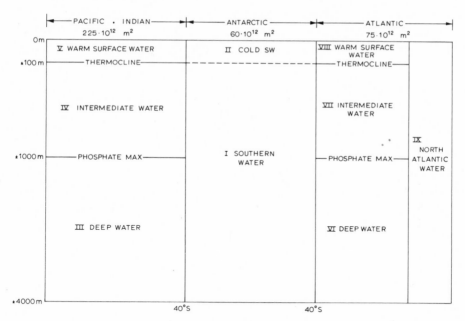

Fig. 8.5. Model for the distribution of phosphate species in the world ocean. (Postma, 1971).

last process is of little importance. His model for water and phosphate transport is given in Fig. 8.5. There are three large oceanic reservoirs: the Pacific–Indian Oceans, the Antarctic Ocean, the circumpolar water of the Southern Ocean, and the Atlantic Ocean. Each has a mixed and deep layer. His mixed-layer reservoirs in the Pacific–Indian Oceans and the Atlantic, which cover about 7% of the ocean surface, are found between $40°S$ and $40°N$. The intermediate water layers IV and VII have their lower boundaries fixed by the phosphate maximum.

The transport of phosphate from one reservoir to another are presumed to take place by advection, the sinking of organic matter and vertical mixing. The principal trajectories are shown in Fig. 8.6. A steady state is assumed with the amounts of phosphate entering and leaving each reservoir being the same. The time constants for removal of phosphate to the sediments are long with respect to mixing processes — hence, there is no need to incorporate a sedimentary reservoir into the model.

The model yields the result that the mixed layers receive 20 mmoles of phosphate per m^2 per year. The primary productivity is estimated to be 36 mmoles per m^2 per year, or about twice the amount received from the deep water. Thus, every phosphate molecule in the upper layers is involved twice in the primary photosynthetic process.

The total amount of upwelling water is estimated to be about $1.8 \cdot 10^{15}$

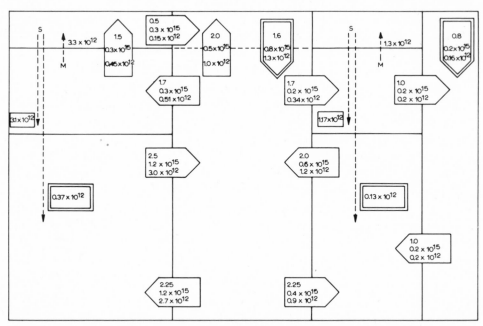

Fig. 8.6. Transport of water and phosphate spécies in Postma's (1971) model of world oceans. Every arrow gives, form top to bottom, the phosphate concentration in mmoles/l; the net amount of water in m³/year and the net amount of phosphate transferred in mmoles/year. S indicates sinking of phosphate with organic matter; M transport by mixing. Steady-state conditions are assumed. Small differences in the balance result from rounding off of numbers.

m³/year. From the model, the net influx into the surface from deeper waters is about $0.3 \cdot 10^{15}$ or, if one includes the Atlantic, perhaps $0.4 \cdot 10^{15}$ m³/year. Even adding a loss from a net evaporation of $0.1 \cdot 10^{15}$ m³/year, we can only in this way account for 25—30% of the upwelling water. Postma suggests that the greater part of the upwelling water must be returned to deeper waters by vertical mixing. The upwelling intensity must be a factor involved with the extensiveness of subtropical convergence regions between 30 and 40° in the Northern and Southern Hemispheres.

Coupling the water transport figures with other nutrients should lead to some most rewarding comparisons with field data. The differences between phosphate and silicate or nitrate distributions should reflect chemical as opposed to physical processes.

8.7. FINALE

One essay on modelling gave new paths for the pursuit of marine chemistry. In 1959 Lars Gunnar Sillén at the First International Oceanographic

Congress in New York put forth his equilibrium model for the sea water system. Three phases (water — sea water; air — the atmosphere; and solids — the sediments) were postulated to be in equilibrium with respect to their chemical constituents. Sillén (1961) argued "It may be worthwhile to try to find out what the true equilibrium would be like and that one might learn something from a comparison with the real world." Sillén recognized that the formulation of such models would be difficult since "Neither the laboratory data on chemical equilibria (needed for the model), nor the geochemical data (for the real system) are always as accurate as one might wish." Nonetheless, the model has proved to be the basis for many novel concepts and major concerns over the past several decades.

Steady-state equilibrium models for the cycling of elements in the major sedimentary cycles, based on Sillén's work, have been constructed by Garrels and MacKenzie (1971). The role of silicate weathering in governing the carbon dioxide contents of the atmosphere and the concentrations of carbonate species in the oceans is developed following the initial idea of Sillén. Although mass balances for silicon during the weathering sequence have been attempted, the sinks for this element in the ocean system are yet to be identified. There appears to be far more silicate species entering the oceans following continental weathering than can be accounted for through removal by biological or inorganic precipitation processes. In a similar way, the magnesium problem has been posed, as a consequence of the equilibrium model. Magnesium is entering the oceans in amounts greater than can be accounted for by removal processes.

One of the challenges to future researches in marine chemistry, derived from the equilibrium models, involves their extension to the deep sea, waters at lower temperatures and higher hydrostatic pressures. Sillén's pioneering work involved a temperature of $25°C$ and a pressure of 1 atm. Waters at abyssal depths, out of contact with the atmosphere or sediments for hundreds of years perhaps, may more closely approach equilibrium speciation of their components than surface waters with their intense biological activity. In addition, with the advent of the interstitial-water samplers previously mentioned, the approaches to equilibrium states of the species in sedimentary pore waters is amenable to study.

REFERENCES

Adam, Y.A., 1973. Phosphorus cycle in a shallow sea pond. Manuscript presented at NATO Science Committee Conference on Modelling of Marine Systems, Ofir, Portugal, 1973.
Barnes, R.O., 1973. *Noble Gas Concentrations in the Pore Fluids of Marine Sediments*

and the Construction of an In Situ Pore Water Sampler. Dissertation, University of California at San Diego, La Jolla, Calif. (unpublished).

Bevenue, A., Hylin, J.W., Kawano, Y. and Kelley, T.W., 1972. Organochlorine pesticide residues in water, sediment, algae and fish, Hawaii — 1970—1971. *Pestic. Monit. Bull.*, 6: 60.

Broecker, W., 1963. Radioisotopes and large-scale oceanic mixing. In: M.N. Hill (Editor), *The Sea*, 2. John Wiley and Sons, New York, N.Y., pp. 88—108.

Clarke, F.W., 1924. The data of geochemistry. *U.S. Geol. Surv. Bull.*, 770: 841 pp.

Conway, E.J., 1943. Mean geochemical data in relation to ocean evolution. *Proc. Ir. Acad.*, 38B: 122—159.

Dyrssen, D. and Wedborg, M., 1974. Equilibrium calculations of the speciation of elements in sea water. In: E.D. Goldberg (Editor), *The Sea*, 5. John Wiley and Sons, New York, N.Y.

Garrels, R.M. and MacKenzie, F.D., 1971. *Evolution of Sedimentary Rocks*. W.W. Norton and Co., New York, N.Y., 397 pp.

Garrels, R.M. and Perry Jr., E.A., 1974. Cycling of carbon, sulfur and oxygen through geologic time. In: E.D. Goldberg (Editor), *The Sea*, 5. John Wiley and Sons, New York, N.Y.

Garrels, R.M. and Thompson, M.E., 1962. A chemical model for sea water at $25°C$ and one atmosphere total pressure. *Am. J. Sci.*, 260: 57—66.

Goldberg, E.D., 1971. Atmospheric dust, the sedimentary cycle and man. Comments on Earth Science. *Geophysics*, 1: 117—132.

Goldberg, E.D. and Arrhenius, G.O.S., 1958. Chemistry of Pacific pelagic sediments. *Geochim. Cosmochim. Acta*, 13: 153—212.

Goldberg, E.D. and Koide, M., 1963. Rates of sediment accumulation in the Indian Ocean. In: J. Geiss and E.D. Goldberg, *Earth Science and Meteoritics*. North-Holland Publishing Company, Amsterdam, pp. 90—102.

Goldschmidt, V.M., 1933. Grundlagen der quantitativen Geochemie. *Fortschr. Mineral., Kristallogr. Petrogr.*, 17: 112—156.

Harvey, G.R., Bowen, V.T., Backus, R.H. and Grice, G.D., 1972. Chlorinated hydrocarbons in open-ocean Atlantic organisms. In: D. Dyrssen and D. Jagner (Editors), *The Changing Chemistry of the Oceans*. Almquist and Wiksell, Stockholm, pp. 177—186.

Harvey, G.R., Steinhauer, W.G. and Teal, J.M., 1973. Polychlorobiphenyls in North Atlantic Ocean Water. *Science*, 180: 643—644.

Horn, M.K. and Adams, J.A.S., 1966. Computer derived geochemical balances and element abundances. *Geochim. Cosmochim. Acta*, 30: 279—297.

Judson, S., 1968. Erosion of the land. *Am. Sci.*, 56: 356—374.

Kester, D.R. and Pytkowicz, R.M., 1968. Sodium, magnesium and calcium sulfate ion-pairs in sea water at $25°C$. *Limnol. Oceanogr.*, 14: 686—692.

Koczy, F.F., 1958. Natural radium as a tracer in the ocean. *Proc. Int. Conf. Peaceful Uses of Atomic Energy, 2nd*, 18 : 351—357.

Liss, P.S., 1973. Processes of gas exchange across an air—water interface. *Deep-Sea Res.*, 20: 221—234.

Liss, P.S. and Slater, P.G., 1973. Use of a two layer model to estimate the flux of various gases across the air—sea interface. Paper submitted to NATO Science Committee Conference on Modelling of Marine Systems, Ofir, Portugal, 1973.

Machta, L., 1972. The role of the oceans and biosphere in the carbon dioxide cycle. In: D. Dyrssen and D. Jagner (Editors), *The Changing Chemistry of the Oceans*. Almquist and Wiksell, Stockholm, pp. 121—145.

MacIntyre, F., 1974. Chemical fractionation and sea-surface microlayer processes. In: E.D. Goldberg (Editor), *The Sea*, 5. John Wiley and Sons, New York, N.Y.

Menzel, D.W., 1974. Primary productivity, dissolved and particulate organic matter and the sites of oxidation of organic matter. In: E.D. Goldberg (Editor), *The Sea*, 5. John Wiley and Sons, New York, N.Y.

Nisbet, I.C.T. and Sarofim, A.F., 1972. Rates and routes of transport of PCBs in the environment. *Environ. Health Perspect.*, 1: 21—38.

Postma, H., 1971. Distribution of nutrients in the sea and the oceanic nutrient cycle. In: J.D. Costlow (Editor), *Fertility of the Sea*, 2. Gordon and Breach, New York, N.Y., pp. 337—349.

Risebrough, R.W., Huggett, R.J., Griffin, J.J. and Goldberg, E.D., 1968. Pesticides: Transatlantic movements in the Northeast Trades. *Science*, 159: 1233—1236.

Sayles, F.L., Wilson, T.R.S., Hume, D.N. and Manglesdorf Jr., P.C., 1973. In situ sampler for marine sedimentary pore waters: evidence for potassium depletion and calcium enrichment. *Science*, 181: 154—156.

Sillén, L.G., 1961. The physical chemistry of sea water. In: M. Sears (Editor), *Oceanography*. Am. Assoc. Adv. Sci., Washington, D.C., pp. 549—581.

Tatsumoto, M. and Patterson, C.C., 1963. The concentration of common lead in sea water. In: J. Geiss and E.D. Goldberg, *Earth Science and Meteoritics*. North-Holland Publishing Co., Amsterdam, pp. 91—102.

Tzur, Y., 1971. Interstitial diffusion and advection of solute in accumulating sediments. *J. Geophys. Res.*, 76: 4208—4211.

Wickman, F.E., 1954. The total amounts of sediments and the composition of the "Average Igneous Rock". *Geochim. Cosmochim. Acta*, 5: 97—110.

Woodwell, G.M., Craig, P.P. and Johnson, H.A., 1971. DDT in the biosphere: where does it go? *Science*, 174: 1101.

CHAPTER 9

BIOLOGICAL MODELLING I

Richard C. Dugdale

9.1. INTRODUCTION

The subject of "biological simulation" implies an area so broad that it is necessary first to delimit the specific objectives and the amount of historical material to be included in this communication. Here, biological simulation models are taken to mean those attempts to build quantitative descriptions of marine organisms or systems at a level of realism and detail that dictates the use of modern electronic computers. This definition has the effect of eliminating discussion of earlier modelling attempts in which so many simplifications were made to allow analytical solutions that the results were usually superficial or of limited applicability. These earlier efforts to understand marine productivity have been discussed by Patten (1968). Present modelling efforts are of course a logical outgrowth of previous work but the emphasis of this paper will be to present and assess the direction of their evolution now. The models discussed are generally of very recent origin and often are published only in the gray literature. No attempt has been made to include all recent models dealing with marine biological systems. Instead, models are selected and discussed for a variety of approaches, with examples taken from terrestrial ecology and other branches of biology where appropriate.

To facilitate organization of the discussion, models are considered to belong somewhere in the following hierarchy:

(1) Ecosystem models   (3) Population models
(2) Productivity models   (4) Process models

Hard and fast definitions of these categories are not required for our purposes. However, it will be recognized that they are listed more or less in order of decreasing complexity. It will be instructive to begin by discussing the most complex category so that the various interrelationships become clear at the outset. The reader should remain aware constantly that the variety of models reflects in part the purposes for which they were constructed.

9.2. EXISTING MODELS

Ecosystem models

This most complex level of model can be distinguished from the next lower level, productivity models, only by degree of complexity; obviously a difficult basis on which to make a distinction and one very much prejudiced by the modeller's point of view. However, at the present time working models of marine ecosystems containing sufficient detail to allow their ready placement in this category do not appear to exist. The same is probably true for terrestrial ecosystem modelling. However, under the International Biological Program, considerable momentum has been imparted to activities leading toward the completion of a number of terrestrial ecosystem models and at least one marine ecosystem model. The development of these models will be discussed in a later section.

Productivity models

These models usually are based on earlier work of Riley and Steele (Riley, 1965; Steele, 1962). In recent models, equations are written as first-order differential equations and later are converted to difference form for solution on digital computers using various numerical techniques. Sets of equations are required for: (1) growth rate of phytoplankton as a function of incoming radiant energy and a limiting nutrient; (2) ambient radiant energy intensities at physiological wavelengths; (3) limiting nutrient concentration as a function of uptake, regeneration, and supply by mixing of surface waters with deep water; (4) loss rates for the phytoplankton, from grazing by zooplankton and fish, from natural mortality, sinking and mixing; and (5) activities of zooplankton and higher trophic levels (these terms may allow for increase in biomass or be held fixed when short periods of time are to be simulated).

The aim of the investigators who build such models often is to simulate the seasonal cycles of phytoplankton standing crop and primary productivity, along with the seasonal levels of the limiting nutrient and zooplankton populations. By manipulation of the model parameters some measure of agreement with the observed pattern of the state variables usually is achieved. Whether such agreement assures the validity of the models is another question to which we will return later.

Spatial aspects

The search for simplification as an aid to obtaining analytical solutions was perhaps pursued most vigorously in the past in the hydrodynamics portions of models. At the most-simplified end of the spatial spectrum, Riley

(1965) introduced the two-layered model, with a homogeneously mixed euphotic zone underlain by a layer with constant nutrient concentration and containing no phytoplankton, i.e. an infinite nutrient source and infinite phytoplankton sink. A mixing coefficient, m, can then be used to compute the transport of nutrient into the euphotic layer and the transport of phytoplankton into the lower layer. For example, the flux of a limiting nutrient can be expressed as:

$$\frac{dN}{dt} = -m(N_U - N_L)$$

where N_U is the concentration of nutrient in the upper layer; N_L that of the lower layer; and m is a mixing coefficient with units of t^{-1}. Lateral homogeneity is assumed, avoiding the necessity for formulation of horizontal advective and diffusive transports. Lassen and Nielsen (1972) are the most recent authors to exploit this approach in constructing a model of the phytoplankton cycles in the North Sea. Steele (1972) has also used this hydrodynamic base for a model designed to study some details of zooplankton feeding and reproduction.

In their effort to simulate the production processes in water upwelled along the Equator in the Pacific Ocean, Vinogradov et al. (1972) adopted the hydrodynamic framework of a multi-layered parcel of water retaining constant shape as it drifts for up to 100 days toward the oligotrophic regions. A lagrangian coordinate system is imposed, apparently with the assumption of negligible horizontal gradients. The vertical space extends from the surface to 200 m and is divided into 10-m segments. The transport of nutrient is computed as a function of the concentration gradient between two adjacent layers and of a diffusion coefficient β. β is assigned one of three possible values depending upon the location of the layer relative to the thermocline depth. Thermocline depth is allowed to increase at a fairly high rate as a linear function of time for the first 50 days, and at a lower rate from that time on. The resulting thermocline depths range from 10 m to 150 m and presumably are based upon observations from the region. Below 200 m the usual infinite, unvarying source of high nutrient deep water is assumed. Although the assumption of zero advective transfers and of zero diffusion terms in the horizontal direction are likely to be damaging especially in the actively upwelling regions close to the origin of the simulation, the introduction of vertical structure is an important step in simulating some of the noteworthy biological features of the more oligotrophic oceanic regions. This model could be classified also as a transitional ecosystem model rather than a production model, since it has considerable complexity built into it.

In another attempt to reproduce the observed spatial patterns of phytoplankton and nutrients under upwelling conditions, Walsh and Dugdale

(1971) constructed a two-layered, one-dimensional model. Five spatial blocks were laid out along the axis of an actual upwelling plume located by drogue studies at the site along the south coast of Peru. The blocks, 11 km × 11 km and 10 m deep, are assumed for the computation to be in a straight line and contiguous. Longshore transport is assumed to be negligible, but offshore transport and upwelling at variable rates decreasing offshore are allowed, as are diffusive transports in the lateral dimension. With tuning, this model reproduced the major observed biological and nutrient features of the plume. Previous attempts to obtain material balances from data taken in the same area while following a parachute drogue for five days had failed (Ryther et al., 1971). The introduction of a more realistic hydrodynamic framework into a simulation model consequently may be said to have allowed some new insight into the productivity processes taking place in a highly dynamic ecosystem.

The model described by O'Brien and Wroblewski (1972) lies at the hydro-dynamically sophisticated end of the spectrum of computer simulation models of marine productivity. These authors added biological terms to a previously developed model of the circulation on the continental shelf along the West Florida coast (Hsueh and O'Brien, 1971). The model is in two dimensions, the offshore direction and z; no longshore variation is permitted. A grid of 41 boxes in the x dimension and 82 in the z dimension forms the computational basis; each box is 2.5 m deep and 5 km in width. Vertical circulation is driven by current-induced upwelling as a southbound current meanders onto the continental shelf; bottom topography is not allowed to vary.

A single-layered, two-dimensional model, in the x and y directions, was constructed by Dugdale and Whitledge (1970) for phytoplankton growth around a marine sewage outfall. The hydrodynamic base was borrowed from Brooks (1959). Flow was allowed in one direction, past a sewage diffuser. Diffusion was allowed in the two lateral dimensions. Ten blocks were arranged normal to the direction of flow and 20 in the downstream direction. The blocks were 300 m × 300 m and 10 m in depth. A similar model with a more sophisticated hydrodynamic base has been made by Hendricks (1973) in conjunction with studies of the San Diego, California, Point Loma sewage outfall. The configuration of boxes is similar to that of the Walsh and Dugdale (1971) Peru model, with a series of boxes aligned with the flow path, in this case following initial diffusion of the sewage.

The point to be made from the array of models presented above is that there is a strong trend toward spatial or multi-dimensional models. Single-point type models can be expanded to multi-dimensional models by inserting the required set of equations into each block, computing the changes in an interval of time and then allowing diffusion and advective transport to take place between that block and adjacent blocks. The method used to generate the advective and diffusion terms varies from simple definition of a steady-state restricted current field, for example in Dugdale and Whitledge (1970), to a steady-state velocity field generated from hydrodynamic equations, for example in O'Brien and Wroblewski (1972).

Choice of units

Biological modellers have traditionally chosen to carry all biological elements of the model in a single unit, usually energy, carbon, or a primary nutrient, that is, phosphorus or nitrogen. However, since most models include a nutrient term placing limits on photosynthesis, conversion factors must be used, for example from carbon to nitrogen. Although carbon and energy units are readily convertible, carbon to nutrient ratios in phytoplankton cells are not constant, especially under conditions of transition from high to low productivity. The models discussed up to this point have all used a single-unit basis. Lassen and Nielsen (1972) use a "phytoplankton unit" equal to 1 mg; Vinogradov et al. (1972) carry state variables in energy units; Steele (1972) uses carbon; while the others discussed use the limiting nutrient approach, in these cases, nitrogen. At times, the coupling between nutrient uptake, photosynthesis and growth is rather loose, and as models become more detailed these elements of the models may have to be accounted for separately.

Phytoplankton terms

The terms affecting growth of the phytoplankton population are indicated in the following word equation:

$$\frac{\text{change of phytoplankton}}{\text{time}} = \text{growth} - \text{respiration} - \text{sinking} - \text{grazing} + \text{advection} + \text{diffusion} - \text{excretion} - \text{natural mortality}$$

Advection and *diffusion* are taken care of automatically with the definition of the hydrodynamic framework and need not be discussed further. *Grazing* will be considered in the zooplankton equation.

Growth. Put in the simplest terms, the growth of individual algal cells in the sea is under the control of light and nutrients. The gradients of these two parameters are, however, in opposite directions, light decreasing and nutrients normally increasing in the downwards direction. The form of the curve for light intensity versus photosynthesis is reasonably well known, taking the form of a saturation curve with an inhibition of photosynthesis occurring at high light intensities. A maximum photosynthetic rate and thus the maximum growth rate for the phytoplankton cells must be specified along with constants determining the shape of the curve. Both Riley and Steele have in the past used a linear reduction of this maximum photosynthetic rate as a function of phosphate concentration below some threshold level, with the effect of producing an interaction term for nutrients and photosynthesis. From another point of view, either light or nutrient concentration may be considered to limit the existing growth rate. Vinogradov et al. (1972) have adopted this latter approach. Their phytoplankton growth rate expression may be extracted from their equation:

individual algal growth rate = $\min[\phi(e), N, \epsilon_p]$

where $\phi(e)$ is the effect of light on photosynthesis and is developed from Ryther (1956); N is the concentration of nutrient; and ϵ_p is the maximum specific growth rate (the maximum rate of photosynthesis per unit biomass of phytoplankton). A term defining the intensity of light at depth is required in such a model, including the effect of the phytoplankton and other particles.

The linear nutrient term of earlier models has been replaced generally by a saturation curve shown by MacIsaac and Dugdale (1969) to govern the uptake of limiting nutrient by phytoplankton under some circumstances:

$$V_N = N \; \frac{V_{max}}{N + k_t}$$

where V_N = rate of uptake of limiting nutrient; V_{max} = the maximum current uptake rate of the limiting nutrient at saturating nutrient concentration; N = concentration of limiting nutrient; k_t = the Michaelis constant, the value of N at which $V = V_{max}/2$.

This formulation is known as the Michaelis-Menten or Monod expression and is used by Dugdale and Whitledge (1970), Walsh and Dugdale (1971), Hendricks (1973), O'Brien and Wroblewski (1972), and Steele (1972). There is little question that this form of the nutrient term is to be preferred at present, though it should not be accepted as definitive. For example a modification made by Lassen and Nielsen (1972) results in an S-shaped curve, with implications that at very low nutrient concentrations little nutrient is taken up. The nutrient term then is used to reduce the maximum photosynthetic rate in the Lassen and Nielsen model, by direct multiplication.

In two-layered models no vertical gradients exist in the euphotic layer. Consequently the rate of mixing of nutrients from below becomes the controlling factor, and an unmodified Michaelis-Menten term can be used for the rate of growth, a method adopted by Steele (1972) to build a model to investigate control mechanisms in the marine trophic system.

Sinking. The loss of phytoplankton by sinking from the euphotic zone is sometimes handled directly by the specification of a mean sinking rate, an approach used generally by both Riley and Steele, but eliminated in Steele (1972), and ignored by Walsh and Dugdale (1971). Hendricks (1973) and Lassen and Nielsen (1972) include the term. Although mathematically simple when a constant value is used, few reliable measurements on natural populations have been made and the investigator is forced to choose a value having little scientific basis.

Another approach to sinking losses is that used by Vinogradov et al. (1972) setting up a term for detritus production, composed of unassimilated food. Neither of these approaches to sinking used singly can be considered satisfactory since both losses occur and perhaps, in some cases, up to half of the phytoplankton losses must be handled in the model in a way in which they may not occur in nature.

Respiratory and excretion losses. Considerable variation exists in the way these losses are treated in productivity models. The models built on energy or carbon bases tend to be more concerned with these terms. Vinogradov et al. (1972) specify that 30% of the phytoplankton production is excreted as dissolved organic matter. A proportional amount of nutrient is carried as a loss by assuming a constant 10% of phytoplankton organic matter to be nutrient. O'Brien and Wroblewski (1972) also treat the excretion loss along with a number of others as a constant proportion of the population. Hendricks (1973) uses a respiratory loss term, adding temperature as a linear term. The assumed equivalence between carbon and nutrient losses encompasses problems that will be discussed later.

Zooplankton terms

In words, the zooplankton term can be written:

$$\frac{\text{change of zooplankton}}{\text{time}} = \text{growth} - \text{mortality} - \text{predation}$$

Growth. This term is the result of grazing expressed as intake of energy, carbon, or nutrient, and respiration and excretion losses. As will be seen later, much more complex formulations are possible. Lassen and Nielsen (1972) use the Riley form in which the grazing rate is the product of the grazing function, usually called a filtration rate, the concentration of zooplankton, and the concentration of phytoplankton:

$$\frac{dP}{dt} = -gZP$$

where g = filtration rate; Z = zooplankton concentration; and P = phytoplankton concentration. From this expression an infinitely high grazing rate

is implied at infinite phytoplankton levels. The sewage outfall model of
Hendricks (1973) uses the same function with fixed zooplankton population.

The linear formulation largely has been replaced by expressions yielding
saturation curves similar to those given by the Michaelis-Menten expression
for nutrient uptake. Steele (1972) uses an expression of the form:

$$\frac{dP}{dt} = C(P-P_1)/(D+P)$$

where C = the maximum grazing rate; P = the phytoplankton concentration;
and P_1 = the threshold level of phytoplankton at which the grazing rate falls
to zero. The validity and interpretation of this threshold feeding level is
currently debated and may be a point of considerable importance. If P_1 is
equal to zero, D is akin to k_t in the Michaelis-Menten expression, i.e., the
phytoplankton concentration at which the feeding rate is reduced to half the
maximum value. Walsh and Dugdale (1971) use the non-linear expression,
inserting a threshold, however:

$$\frac{dP}{dt} = -g(Z)(P-P_1)$$

O'Brien and Wroblewski (1972) begin with a form of the Ivlev equation:

$$I = E_z(1 - e^{-d_p(P-P_1)})$$

where E_z = maximum rate of ingestion; I = rate of ingestion per unit concentration of grazer; d_p defines the shape of the curve with $1/d_p$ being the
concentration of P at which $I = 2/3\ E_z$, providing that $P_1 = 0$, a simplification
made by the authors in running their model.

The Vinogradov et al. (1972) model becomes complex above the phytoplankton level, with eight additional trophic divisions. In fact, this model
approaches an ecosystem level model as defined roughly in the introduction
to this paper, lacking detail primarily in the hydrodynamic base. Bacteria
and detritus compartments are provided and in the levels above, the compartments are small-sized herbivorous zooplankton, omnivorous zooplankton, and carnivorous zooplankton. Small-sized herbivores are defined as
those of size less than 1.0 mm, etc. The general scheme for arriving at the
ingestion rate for a given trophic level is: (1) to compute the daily requirement of food of a trophic level as a proportion of its own biomass; (2) to
apportion this requirement among the trophic levels providing food in direct
proportion to the biomasses of each of these providing levels; and (3) to
reduce this maximum requirement to the real daily intake by an Ivlev (1961)
function having the effect of decreasing the actual feeding rate of the trophic

level as the rate of feeding on the trophic level diminishes. The function used to perform this multiplication is:

$$\sigma_i = 1 - e^{-\xi k}$$

In this paper ξ is undefined and taken to be equal to 1 and k = specific rate of feeding by other trophic levels on the level under consideration. Laboratory information on assimilation efficiencies are used to develop the maximum daily rations.

In the Vinogradov et al. (1972) model, provision is made for grazing in the 0—50-m layer by small herbivore and higher trophic-level elements residing in the 50—200-m level, i.e. for vertical migration of zooplankton.

Respiration and excretion. The amounts of carbon and nutrient ingested as phytoplankton by herbivores are partitioned into growth (including reproductive products) fecal pellets, excretion, and in the case of carbon, into respiratory losses. The manner in which modellers handle this problem varies. Lassen and Nielsen (1972) make no provision for excretion of nutrient and carry no compartment for detritus. The latter condition is of no importance for the purpose of their model, but the omission of an excretion term to allow a return of phosphate to the nutrient pool is probably a factor in their difficulty in simulating the phosphate cycle in the North Sea.

Steele (1972) computes a respiration rate dependent on feeding rate and converts this number by a carbon/nitrogen ratio to give an excretion rate, partitioning the excreted nitrogen between upper and lower layers on the basis of time spent in the two places as a result of vertical migrations of the herbivores. The biomass increase is converted into copepods weighing 20 μg initially and 100 μg as adults. After reaching the latter value, increase in biomass is partitioned into egg production. The size of the next generation is computed from a proportion of the carbon in eggs. Fecal pellet production is not included in the zooplankton equation, nor is sinking of phytoplankton, so that the entire losses of phytoplankton not assimilated by the zooplankton are channelled through zooplankton respiration and excretion.

Walsh and Dugdale (1971) use zooplankton at levels fixed in time but varying in space to investigate the phytoplankton dynamics of a Peru upwelling system, making it unnecessary to keep a material balance in the zooplankton trophic level. The effect of excretion is inserted as an increased nutrient concentration in the down-plume or offshore direction, with a diurnal variation.

Excretion of nutrient by zooplankton is obtained as a constant proportion of ingestion by O'Brien and Wroblewski (1972). However, these authors do not provide for production of feces by zooplankton although they allow zooplankton to feed on detritus produced as a constant factor times the amount of zooplankton, a term accounting for natural mortality of zoo-plankton.

Hendricks (1973) assigns all unassimilated nutrient back into the nutrient pool. With constant grazing rate and assimilation coefficient, regenerated nutrient is a constant fraction of the ingested phytoplankton.

Vinogradov et al. (1972) sum the respiratory losses for each depth interval and each trophic level considered. Respiration is taken to be a constant proportion of the biomass for a given trophic level. In a similar manner, the detritus component is produced from undigested food at each trophic level.

Predation. Models often are terminated at the herbivore level especially if the objective is to predict phytoplankton dynamics, as in Walsh and Dugdale (1971), Lassen and Nielsen (1972), and Hendricks (1973). Steele's (1972) model was designed specifically to investigate the effects of different strate-gies of predation, the feeding of carnivores or omnivores on herbivores. He uses two alternative patterns. In the first, predation on numbers of zooplank-ton is a fixed proportion of the population at any weight or size. In the other, predation is dependent upon biomass:

$$ZW = -G(Z - Z_1)(W - W_1)/(H + ZW)$$

which is a Michaelis-Menten function with weight and number thresholds built in, that is, if numbers decrease to Z_1, or if weight decreases to W_1, predation decreases to zero.

The technique used by Vinogradov et al. (1972) to assign feeding rates between various trophic levels has already been described. Essentially these values are determined on the basis of relative abundance and no thresholds are used. O'Brien and Wroblewski (1972) include grazing and predation by fish in their model. Grazing on zooplankton is formulated as:

$$\frac{dP}{dt} = -\phi F\left(\frac{\theta P}{\theta P + Z}\right)$$

where F = the concentration of fish and sets a maximum daily ration as a proportion of fish biomass; and θ = the gill raker efficiency. The effect of Z in the denominator of the parenthetical element is to reduce predation on phytoplankton when zooplankton are present. However, in the absence of zooplankton the feeding rate on phytoplankton becomes independent of phytoplankton concentration, depending only upon the concentration of fish. In similar fashion, fish predation on zooplankton is formulated:

$$\frac{dZ}{dt} = -\phi F\left(\frac{Z}{Z + \theta P}\right)$$

The intent is again to weight predation as a function of phytoplankton abundance. The same problem as before arises at low phytoplankton concentrations. In this model fish biomass is held constant, with all ingested nutrient excreted and sent back to the dissolved nutrient pool.

Evaluation of the current status of productivity models

Each of the six models considered up to this point was constructed with a set of specific objectives in mind. The extent to which these objectives were reached varied. Probably none of the authors would at this time claim much in the way of predictive capability for his model. The purpose of reviewing this set of models was to assess the various approaches being used currently to build models and to try to make some conclusions about the state of the art. All of the models had, of course, considerable heuristic value.

The hydrodynamic bases vary from the simplistic two-layered, single-point system to three-dimensional models incorporating considerable hydrodynamic theory. Biological modelling and hydrodynamic modelling have converged upon the compartmental block or grid point representation and method of handling spatial models, with the result that the state-of-the-art in the area of handling advective and diffusive transfers of water is well advanced for steady-state models at least. A three-dimensional simulation system called AUGUR, written by Walsh, Bass, and Morishima for the CDC 6400 computer, allows specification of the advective terms, the size and number of blocks, and the specification of bottom topography and variable-sized blocks at boundaries. With the ability to handle up to 33,000 blocks and up to 40 state variables, AUGUR provides a powerful tool for marine ecosystem modelling. To make this generalized program approach easier to use and more attractive to biological modellers, a library of advective regimes should be developed both from measurements in the field and from the output of hydrodynamic simulation models. The biological modeller then could test his ideas in a spatial setting almost as easily as by using gross simplifications that may be self-defeating.

Nutrients, phytoplankton and units. The Michaelis-Menten expression has become the standard expression for nutrient uptake within the last five years, partly as a result of a theoretical treatment (Dugdale, 1967) and a growing body of experimental evidence that the expression was widely applicable to marine phytoplankton and to phytoplankton populations in the sea (Eppley and Thomas, 1969). However, more detailed sub-models for nutrients can be written based on laboratory work with continuous cultures (Conway, 1974; Davis, 1973; and Harrison, 1974) and from additional field observations (MacIsaac and Dugdale, 1972). One sub-model, for nitrate uptake, appears to predict the vertical distribution of nitrate uptake in a eutrophic region (Dugdale and MacIsaac, 1971). The V_{max} for nitrate uptake is set as a function of ambient nitrate concentration with provision for suppression by ambient ammonia. A term for light is included as a Michaelis-Menten expression. An uptake calculated from the nutrient expression is compared with that calculated from the light expression, and the lesser of the two is selected and stored as the value for the time interval and depth indicated. It is necessary to carry both nitrate and ammonium separately in the nutrient sector of the model to use these sub-models. There are other strong reasons for carrying ammonium separately from nitrate since zooplankton excrete nitrogen in the form of ammonium and some urea as well. The production based on nitrate is called new production and that based on ammonium or urea is called regenerated production (Dugdale and Goering, 1967). The ammonium so produced has a regulatory effect on nitrate uptake as it is itself taken up into the phytoplankton. This regulatory effect takes place at very low concentrations of ammonia, less than 1 μgat/l.

In general, it appears that a separate compartment must be set aside for carbon since the ratio of carbon to limiting nutrient varies. For example, photosynthesis may continue at high rates for some time following nutrient depletion. If the investigator is not specifically interested in carbon as a state variable, the effect of light can be carried along with the nutrient as indicated above.

As a result of recent laboratory investigations, it has become clear that V_{max} for nutrient uptake is a somewhat more complicated parameter than had been thought. The subject is too complex to discuss in detail here, and the present assessment is that no additional uncertainties are likely to arise from the concept of a constant V_{max} so long as steady-state solutions are desired without regard to the time sequence of the approach to steady state.

Sinking. The treatment of sinking rates given in the six models considered varies from ignoring the term to assuming some constant rate for the full length of the simulation and for the full spatial scale. In terms of existing theory and field measurements the latter treatment is state-of-the-art, a truly appalling state of affairs. Sinking rate is undoubtedly dependent on time, age, or growth rate in phytoplankton. We have, for example, repeatedly observed a sudden overnight sinking-out of populations of diatoms in the continuous culture laboratory, although at this point no ready explanation is available.

Respiration and excretion losses. The respiration and excretion losses are not treated in an advanced way in any of the models discussed here. Phytoplankton excretion appears to be a function of nutrient limitation, being rather small in healthy, well-nourished phytoplankton (Ignatiades, personal communication, 1973), on the order of 10%. Further, the compounds excreted may not contain much in the way of nitrogen or phosphorus. Respiratory losses are usually ignored as the photosynthetic rate measured with ^{14}C is assumed to lie somewhere between gross and net photosynthesis. Packard et al. (1972) have obtained data on phytoplankton respiration using an enzymatic assay that may be used to assess these loss rates more accurately, but simultaneously, photosynthetic rates must be expressed as gross rates. Hendricks (1973) is the only modeller considered here who included a respiration term in his model. Clearly, the carbon loss portions of the phytoplankton terms in these models are not state-of-the-art. Excretion losses from phytoplankton are not well studied and in this sense the choice is between (1) the assumption of equivalence between carbon loss and nutrient loss, and (2) ignoring the term.

Zooplankton terms. There appears to be general agreement among experimental scientists and modellers that some kind of saturation curve expresses the relationship between phytoplankton concentration and grazing rate. However, considerable controversy exists on the concept of a minimum concentration of phytoplankton at which grazing ceases. Steele's (1972) model was designed to investigate this problem and he obtained evidence that the phytoplankton—herbivore system is unstable without such a threshold unless a very high grazing constant is used. Frost (personal communication, 1973), however, has been unable to confirm the observations of a feeding threshold reported by McAllister (1970) suggesting that considerable work remains to be done in this area. Although Steele's conclusions may prove correct, it seems possible that too few factors have been entered into the model. For example, the results might be different with multi-species systems, a point made in his paper. The modelling is state-of-the-art, but the modelling approaches and the experimental evidence leave room for considerable future advances. Vinogradov et al. (1972) using a single pool for phytoplankton, but a fairly complex feeding structure, reported no difficulties with instabilities, although they use no thresholds in any grazing or predation terms. The length of simulated time, 60 days, was relatively small compared to the iteration interval, 1 day, and it is possible that instabilities might result at longer simulated time.

Respiration and excretion. Excretion is usually computed from respiration values or from ingestion by the application of assimilation efficiencies. By implication regeneration of fecal material is assumed to be instantaneous. The inclusion of specific terms for fecal production accompanied by fecal sinking rates is probably required since the latter term is likely to have a high value, resulting in significant losses of nutrient to lower layers or to the sediment—water interface.

9.3. SUMMARY OF OVERALL STATE-OF-THE-ART IN BIOLOGICAL PRODUCTIVI-
TY SIMULATION

As we have seen from the material presented, the technical problems of programming and of embedding biological models in a variety of hydrodynamic settings are in a reasonable state. As should also be obvious, there is a major problem with the biological sector; the problem that there is so little biology contained in the models! It is reflected, for example, by the search for single time-invariant values of V_{max}, k_t, and sinking rate for algae; in the search for single time-invariant values for grazing, fecal production, and excretion constants for zooplankton, etc. Holling and Ewing (1969) make the point repeatedly that the direction of biological modelling has been so conditioned by the early means available to solve equations, i.e., by the necessity to write differential equations that can be solved analytically, that it has been difficult for biologists to grasp the means now available to them through the use of large digital computers.

There is a series of parameters that are primarily physiological, and these can probably be derived without excessive difficulty from existing knowledge. For example, respiration, excretion, and photosynthesis are processes that fall easily into this category of essentially chemical or biochemical terms. The other category of parameters is that in which a strong element of behavior or dependence upon stage of the life cycle is a characteristic element. For phytoplankton parameters, one may include in the latter category V_{max} (and perhaps k_t) since *Skeletonema costatum*, a diatom, for example has been shown to have a variable V_{max} depending upon the stage of the cycle of sexual reproduction (Davis et al., 1973). The sinking rate appears also to be a function of the stage of the life cycle in diatoms, and when dinoflagellates are considered, the sinking rate may take a negative sign at times of the day, i.e., they may migrate vertically. In the zooplankton sector, a crucial parameter is the grazing rate. At this point, animal behavior becomes an enormously complicating factor. The vertical migration of zooplankton is a complex problem also with impact on the effect of grazing terms in models. The inclusion of threshold concentrations of phytoplankton in the grazing term of two of the models considered here, and the consideration of vertical migration in another, add up to the total content of "zooplankton biology" in the models considered. Undoubtedly this is one reason why many biologists are deeply suspicious of modelling efforts of the variety discussed here.

9.4. POPULATION MODELS

If the community-wide parameters used in most productivity models are considered too gross to adequately represent the ecosystem processes in models, another approach is to build models on the basis of the species components of the ecosystem. One attempt along these lines was described by Efford and Walters (1970). In a cooperative effort under the auspices of the Canadian component of the International Biological Program, a group of 25 students and faculty worked with Drs. Efford and Walters to construct a model of the energy flow in the Marion Lake ecosystem. The model is based on a set of sub-models for mortality, attack rate, food intake, metabolism, growth, and reproduction. The parameters for each sub-model were indexed and stored for use by each sub-model; 24 species eventually were included. Time bases, order of calculation, and physical parameters are provided from a book-keeping sub-model. Calculations were made beginning at the top of the food web. Some of the sub-models were more extensive than others, for example, the grazing and predation functions are derived from the highly advanced models for predation described by Holling and Ewing (1969). The entire approach to the Marion Lake modelling effort bears the stamp of Holling's influence.

Dr. C. Walters, in reporting the preliminary results of the Marion Lake work (Working Conference of the Marine Section, International Biological Program, FAO, Rome, October 1971), indicated that the model perhaps had not performed for predation as well as might have been hoped. Two reasons were reported: (1) species changes occurred at relatively rapid rates, leaving the ecosystem whose state was to be predicted, a different one than that being modelled; and (2) failure of the model to fit valid observational data could usually be traced to spatial problems, the Marion Lake model having no spatial elements.

Species-level simulation models are being constructed in other areas of biology. For example, EPIDEM, a simulator of the early blight of tomatoes and potatoes reported by Waggoner and Horsfall (1969) was designed for the prediction of the success of the disease-producing organism *Alternaria solani* as a function of the life stages of the organism, conidiophores, spores, transport by wind or rain, germination, penetration of host, incubation and expansion of the lesion. The state variables were numbers of individuals at each of the above stages of the life cycle. Input parameters were temperature, light, rainfall, wind and humidity with separate formulations for the effects of these parameters at each life cycle stage. Missing data were obtained experimentally and real data for weather and abundance of the disease at an experimental farm were used to run the model, with considerable success according to the authors. They were, for example, able to explain that the disease was epidemic in Israel as a result of dew and to explain other previously unexplained aspects of the epidemiology of a commercially significant disease. Coulman et al. (1972) report the results of a simulation population model of a fresh-water shrimp, *Hyallela azteca*; all parameters of natural population growth were included in a life table for each of the life stages modelled, an improvement over the Marion Lake model, the latter initially limited by computer storage capacity. The model was built using a large body of field data available for the organism.

9.5. CONCLUSIONS AND FORECASTS

We have seen that the biological content of present marine ecosystem models is unsatisfactory and many improvements remain to be made. The pattern set by the population modellers, using sub-models incorporating biological detail, will undoubtedly be followed. One model, of an upwelling ecosystem, is to be built on this plan. The sub-models are to be built by biologists who are specialists in each subject, using the best laboratory and field data available. The process model will be iterative in the Coastal Upwelling Ecosystems Program between modelling efforts and field efforts. It appears that this technique, which is in a sense building from the opposite direction from the usual, will help to overcome the mathematical biases so apparent in present models. Walters (1970, pers. comm.) considers the involvement of specialists vital to a modelling program. In his words, "...any good field biologist can become as good a simulationist as any mathematician (can)...". Another large-scale ecosystems model (Math. Modelsea, 1972, 1973, 1974) has emerged from the work of a large group of scientists actively involved in modelling the Southern Bight of the North Sea. Goodall (1972) has discussed clearly the philosophical approaches to ecosystem modelling, considering the selection of model hierarchies and the level of lumping. The latter is important since the species level is unreasonably detailed and the herbivore and phytoplankton levels, for example, are too coarse to give reasonable results. It seems possible that both subdivision of trophic levels and the inclusion of spatial aspects may help clarify the issue of the instabilities arising in simple models. In addition to Steele's (1972) discussion reported here, Rosenzweig (1971) and McAllister et al. (1972) have engaged in discussions of ecosystem stability based on two-species models. At some point a distinction must be made between the instabilities resulting from mathematical abstraction and those resulting from real ecosystem processes.

The problem of validation deserves serious consideration, and an encouraging trend to build validation experiments into the modelling process is apparent. Models with enough parameters can usually be tuned to give a number of state variables that match a set of observations. Walters, as mentioned previously, is suspicious of such models and rightly so. A lot of tuning for example went into the Peru model of Walsh and Dugdale (1971) and by inference, one can guess that the same was true of O'Brien and Wroblewski's (1972) model. The procedure is natural and reasonable; however, it raises the question of validation to a more serious level. The matching of observed and predicted standing crops of phytoplankton, herbivores, etc., cannot be taken as proof that the model is correctly simulating the processes in the model since the standing crops are the integrals of the process. Goodall (1972) suggests that each sub-model should be validated experimentally to whatever extent possible. For marine ecosystems models, validation should consist of comparing predicted rates with measured rates during validation cruises, along with comparisons of the concentrations of organisms observed and predicted.

The enormous power of modern large computers gives biologists the opportunity to put our collective detailed knowledge of marine biological processes into an integrating framework. In addition to this integrating of biological detail, the power of computing to give us a spatially inhomogeneous sea will probably prove to be one of the most important factors in the biological oceanography of the current decade, allowing us to come to grips mathematically with the scenes of aggregation we see from the decks of ships and which our intuition tells us must not be ignored in theories of oceanic production processes.

REFERENCES

Brooks, N.H., 1959. Diffusion of sewage effluent in an ocean current. In: E.A. Pearson (Editor), *Waste Disposal in the Marine Environment.* Pergamon Press, New York, N.Y., pp. 246—267.
Conway, H.L., 1974. *The Uptake and Assimilation of Inorganic Nitrogen by Skeletonema costatum (Greve) Cleve.* Dissertation, University of Washington, Seattle, Wash.
Coulman, G.A., Rose, S.R. and Tummala, R.L., 1972. Population modelling: a system's approach. *Science,* 175: 518—521.
Davis, C.O., 1973. *Effects of Changes in Light Intensity and Photoperiod on the Silicate-Limited Continuous Culture of the Marine Diatom Skeletonema costatum (Greve) Cleve.* Dissertation, Dept. of Oceanography, University of Washington, Seattle, Wash., 122 pp.
Davis, C.O., Harrison, P.J. and Dugdale, R.C., 1973. Continuous culture of marine diatoms under silicate limitation. I. Synchronized life-cycle of *Skeletonema costatum. J. Phycol.,* 9: 175—180.
Dugdale, R.C., 1967. Nutrient limitation in the sea: dynamics, identification and significance. *Limnol. Oceanogr.,* 12: 685—695.
Dugdale, R.C. and Goering, J.J., 1967. Uptake of new and regenerated forms of nitrogen in primary productivity. *Limnol. Oceanogr.,* 12: 196—206.

Dugdale, R.C. and MacIsaac, J.J., 1971. A computation model for the uptake of nitrate in the Peru upwelling region. *Invest. Pesq.*, 35: 299—309.

Dugdale, R.C. and Whitledge, T.E., 1970. Computer simulation of phytoplankton growth near a marine sewage outfall. *Rev. Int. Oceanogr. Med.*, XVII: 201—210.

Efford, I.E. and Walters, C.J., 1970. A model for Marion Lake (secondary production processes). *Workshop on Systems Modelling in Ecosystem Studies, Canadian Committee on the International Biological Program.* Institute of Animal Resource Ecology, University of British Columbia, Vancouver, B.C., 1970.

Eppley, R.W. and Thomas, W.H., 1969. Comparison of half-saturation constants for growth and nitrate uptake of marine phytoplankton. *J. Phycol.*, 5: 375—379.

Goodall, D.W., 1972. Building and testing ecosystem models. In: J.N.R. Jeffers (Editor), *Mathematical Models in Ecology.* Blackwell Scientific Publications, London, pp. 173—194.

Harrison, P.J., 1974. *Continuous Culture of the Marine Diatom Skeletonema costatum (Greve) Cleve Under Silicate Limitation.* Dissertation, University of Washington, Seattle, Wash.

Hendricks, T., 1973. The ecology of the Southern California Bight: Implications for water quality management. *Southern Calif. Coastal Water Res. Project* TR104M, Elsegundo, Calif.

Holling, C.S. and Ewing, S., 1969. Blind-man's buff — exploring the response space generated by realistic ecological simulation models. *Proc. Int. Symp. Statistical Ecology, August, 1969.*

Hsueh, Y. and O'Brien, J.J., 1971. Steady coastal upwelling induced by an alongshore current. *J. Phys. Oceanogr.*, 1: 180—186.

Ivlev, V.S., 1961. Experimental Ecology of the Feeding of Fishes. Yale University Press, New Haven, Conn., 302 pp.

Lassen, H. and Nielsen, P.B., 1972. Simple mathematical model for the primary production as a function of the phosphate concentration and incoming solar energy applied to the North Sea. *ICES, Plankton Committee*, CM 1972/L: 6.

MacIsaac, J.J. and Dugdale, R.C., 1969. The kinetics of nitrate and ammonia uptake by natural populations of marine phytoplankton. *Deep-Sea Res.*, 16: 45—57.

MacIsaac, J.J. and Dugdale, R.C., 1972. Interactions of light and inorganic nitrogen in controlling nitrogen uptake in the sea. *Deep-Sea Res.*, 19: 209—232.

Math. Modelsea, 1972. Mathematical models of continental seas, preliminary results concerning the Southern Bight. *ICES, Fisheries Improvement Committee, Hydrography Committee*, CM 1972/E(9): 160 pp.; CM 1973/E(19): 453 pp.; CM 1974/E(29): 472 pp.

McAllister, C.D., 1970. Zooplankton ratios, phytoplankton mortality, and the estimation of marine production. In: J.H. Steele (Editor), *Marine Food Chains.* University of California Press, Berkeley, Calif., pp. 419—457.

McAllister, C.D., LeBrasseur, R.J. and Parsons, T.R., 1972. Stability of enriched aquatic ecosystems. *Science*, 175: 562—564.

O'Brien, J.J. and Wroblewski, J.S., 1972. *An Ecological Model of the Lower Marine Trophic Levels on the Continental Shelf of West Florida.* Technical Report, Geophysical Fluid Dynamics Institute, Florida State University, Tallahassee, Fla., 170 pp.

Packard, T.T., Moore, T., Harmon, D., Devol, A. and King, R., 1972. Respiratory electron transport activity in the euphotic zone plankton of the western Mediterranean Sea, North Atlantic Ocean and the north Pacific Ocean. *CIESM Congress, 23rd, Athens, 1972.*

Patten, B.C., 1968. Mathematical models of plankton production. *Int. Rev. Ges. Hydrobiol.*, 53: 357—408.

Riley, G.A., 1965. A mathematical model of regional variations in plankton. *Limnol. Oceanogr.*, 10 (Suppl.): R202—R215.

Rosenzweig, M.L., 1971. Paradox of enrichment: destabilization of exploitation ecosystems in ecological time. *Science*, 171: 385—387.

Ryther, J.H., 1956. Photosynthesis in the ocean as a function of light intensity. *Limnol. Oceanogr.*, 1: 61—70.

Ryther, J.H., Menzel, D.W., Hulburt, E.M., Lorenzen, C.J. and Corwin, N., 1971. The production and utilization of organic matter in the Peru Coastal Current. *Invest. Pesq.*, 35: 43—59.

Steele, J.H., 1962. Environmental control of photosynthesis in the sea. *Limnol. Oceanogr.*, 7: 137—150.

Steele, J.H., 1972. Factors controlling marine ecosystems. In: D. Dyrssen and D. Jagner (Editors), *The Changing Chemistry of the Oceans. Nobel Symposium 20*. Almquist and Wiksell, Stockholm, pp. 209—221.

Vinogradov, M.E., Menshutkin, V.V. and Shuskina, E.A., 1972. On mathematical simulation of a pelagic ecosystem in tropical waters of the ocean. *Mar. Biol.*, 16: 261—268.

Waggoner, P.E. and Horsfall, J.G., 1969. EPIDEM, a simulator of plant disease written for a computer. *Bull. Conn. Agric. Exp. Station, New Haven*, 698: 80 pp.

Walsh, J.J. and Dugdale, R.C., 1971. Simulation model of the nitrogen flow in the Peruvian upwelling system. *Invest. Pesq.*, 35: 309—330.

CHAPTER 10

BIOLOGICAL MODELLING II

John H. Steele

Some of the general issues that arise in modelling biological systems in the sea are discussed in the contributions by Margalef (1973) and Mortimer (Chapter 11). Many of the detailed problems in producing an adequate simulation are considered by Dugdale. These authors, and many others, stress the interaction between modelling, experimental studies and field observations. The usefulness of any model depends on the quality of the data used as a basis for the postulates and as a test of the output. For example, the "respiration" rate of phytoplankton is an important parameter in most models and yet many of the formulations do not use recent results but are ultimately traceable to Gordon Riley's early work (Riley et al., 1949). Similarly, the problem with the output from most models which are intended to represent the natural environment, is the generally poor fit with observation. Again, the agreement is often no better than that obtained by Riley with his original work in this field. Thus it is necessary to recognise that the main limitations on the application of theoretical simulation lies outwith the simulation.

Strickland (1972), in his review of plankton ecology, suggested that simulation models could be expected to be of use only in the interpretation of controlled experimental systems. Whether or not this turns out to be so, experimental ecosystems can provide observations of better statistical quality than may be obtained in the field. Several groups are developing such systems either in tanks ashore [1,2]*, or in floating containers [3]. These semi-controlled, semi-natural systems may be considered to combine the best or the worst of the two worlds of experiment and observation but, until there is improvement in the collection of data at sea, they may provide the best practical test of theories.

Turning to the models themselves, Mortimer defines two objectives: (1) useful predictive capability, and (2) insight into the workings of a system. Very roughly this might be thought to correspond to an empirical, statistical

---

* Numbers in brackets refer to separate list of papers submitted to the conference (see p. 215).

approach and to dynamic modelling. It would be practically valuable if empirical methods gave useful predictions even if they did not provide any deep understanding of biological mechanisms. It would be equally valuable if dynamic models based on recent physiological, biochemical and behavioural research could indicate new and potentially significant interaction at the population level, even if they gave only the broadest of fits to observation.

In discussing these possibilities I have used the papers presented at the meeting as being representative of the present state of the art.

Ulanowicz et al. [1], Mountford [4] and Pichot [5] all used statistical methods to obtain fits to their data. Ulanowicz et al., in their study of 800 l plankton microcosms, measured seventeen parameters but only three showed discernible trends and so could be used for an analysis of rates of change. These three were nitrogen (NO_3 and NO_2), chlorophyll a and zooplankton biomass. The authors assumed that the rates of change in these three variables can be expressed as general quadratic functions of the three variables. They derived coefficients for the significant terms by minimising the errors.

Mountford [4] collected data on sixteen variables from two bays and, by a multiple regression technique, determined those that were significant in relation to variations of gross photosynthesis as measured by the O_2 technique. Three physical factors were significant, temperature, salinity and tide, but only two biological variables, chlorophyll a and phytoflagellate counts. For each bay the photosynthesis calculated from the regressions over one year gave a reasonably good fit to the data for this year, but the fit became much worse when the regression was used to predict photosynthesis in the following year. If the relation for one bay was used to predict events in the other bay, the fit was not satisfactory.

In a study of an enclosed body of water in Belgium, Pichot [5] used a similar statistical method to determine the form of the equations and the values of the coefficients. Again, from a larger number of variables, there emerged three interacting parameters: nitrate, chlorophyll a and bacterial numbers, with the broad trends but not the details being described.

As Mountford pointed out, in this statistical type of approach, where relations are derived from data and then re-applied to the same data, the results are calculative not predictive. Mountford's results indicate that the general predictive capability may not be very great. In each case described here, there was a large reduction from the number of parameters measured to those found to be significant on some statistical basis. It is interesting that in all these studies chlorophyll a was significant and in two nitrate was significant. This gives some justification for the fact that these are the two most commonly measured parameters. There is also the somewhat negative implication that a very small number of variables is required to describe a system.

In these empirical approaches, the significant variables are deduced from the data, and the functional relations are predetermined by the method of analysis. In dynamic models both the variables and their functional relations are chosen beforehand on the basis of experimental work. Thus, it would appear that dynamic models are more predictive. However, there is usually a wide choice of assumptions and, for each assumption, a range of values for the coefficients involved. This arises from the generally vague formulation of functional relations that can be derived from the literature. In some areas, such as nutrient kinetics, the functional relation for individual species are well described (see Dugdale, Chapter 9), but there are problems when this must be applied to mixed populations where each species may have different coefficients. In other cases, such as zooplankton metabolism, the form of the relation is not established. This means that each systemization is not really one model but several. Dugdale has used the term "tuning" to describe the various adjustments that are usually made to produce reasonable looking outputs. The effects of variation in coefficients can be studied by sensitivity analysis, well described by O'Brien and Wroblewski [6]. This can provide an indication of the significance of individual coefficients in determining the output from a model.

There are a large number of examples of this approach which really started with Riley et al. (1949). Dugdale (Chapter 9) describes some of the more recent work. In the conference, two papers applied dynamical modelling to particular areas. O'Connor et al. [7] modelled an estuarine area with complex flow patterns requiring considerable input of information on tidal and advective flows. The three state variables were nitrogen, chlorophyll and zooplankton biomass. The data to which these parameters are compared showed significant differences between a year of low and a year of high river flow. The model discriminates adequately between these two cases. O'Connor et al. use this model to predict the effects of a threefold increase in nutrient input to the area. The predicted levels of the state variables are well outside the range of present-day values even though, as usual, there is considerable variability in the data. This indicates the *possible* consequences of such a nutrient increase. It also emphasizes the dilemma in the practical application of such models. Because of variability in data, only large extrapolations are likely to give significantly different predictions. On the other hand, such large changes in input might produce basic changes in functional responses such as nutrient uptake, or grazing rate and so invalidate the model being used. The deleterious effects of eutrophication appear to result from the breakdown of one ecosystem and the appearance of another, rather than being explicable in terms of a range of states within one ecosystem with a relatively simple set of characteristics which we can model at present (Steele, 1974).

Buckingham [8] modelled the changes of nitrogen, chlorophyll and zoo-plankton biomass in the Gulf of St. Lawrence by following the changes in a parcel of water as it flowed down the Gulf, using the same general type of model as O'Connor et al. (or as described by Dugdale). Once more, the variability of the data permits only the general conclusion that the predicted levels and changes downstream for each variable are of the right order.

Both these models (and many others) have one thing in common: the state variables are nitrogen, chlorophyll and zooplankton biomass. These are considered to describe the main features of the upper, euphotic, layers of these "internal seas". For a not dissimilar area, the Baltic, Jansson [9] gave a description of a system which differs from the others presented at the meeting. His concern was the hypolimnion, the region below the euphotic zone where the regenerative processes occur. Although the physical aspects are similar to upper-layer processes, involving advection from the North Sea and mixing between the upper and lower layer, his state variables for the eco-system were oxygen, phosphate, organic matter and bacteria. The simulation method used an analogue computer and the equations, built up from a flow diagram, are essentially multiplicative combinations of the variables, similar to the "empirical" approach but involving terms up to cubic. A further difference is that, whereas upper-layer models usually deal with rapidly changing systems over periods of weeks or months, the problem in the Baltic and Jansson's simulation spans periods of tens of years during which there are upward and downward trends in phosphate and oxygen respectively (Jansson, 1972). Apart from the details of this work, it emphasizes the fact that models of the upper layer usually have little or no structuring of the deep-water regenerative processes. Yet, in nature, the processes in these two layers are fairly closely coupled. The problem is partly ecological since we have even less information on rate processes in deeper water and in the bottom than we have for the upper layer; partly analytical since, although some of the coupling will be on short time scales, a major part, as Jansson has shown, is on much larger time scales and yet can produce sudden "switches" in structure (in his case from aerobic to anaerobic metabolism).

This discussion, so far, appears to bring out two points. First, the empirical approach does not appear to have any greater predictive capability than can be obtained from dynamic modelling. This has little to do with the kind of method used but is a consequence of the quality of the data. Limitations are imposed by the complexity of the input based on experimental work and required for a full description, and by the variability of the field data used to test the analytical or numerical theory. Even for phytoplankton production where most is known (and was reviewed by Eppley [10]) it was maintained by Fee [11] that, for lakes, a purely empirical description of the data is the only feasible method.

It remains true that most of the intricacies in response described by Eppley are not incorporated in models, and very simple relations are used for factors such as plant respiration rate where recent evidence suggests complex dependence on other parameters in the plant's metabolism. The second common feature of the models of events in the upper layers of the sea is that the number of state variables is usually small and is usually the same whatever the area studied. These variables are nitrogen, chlorophyll and zooplankton biomass. This gives an impression of generality which may well be spurious, resulting from the inadequacies of the test data. One example of the limitation set by the data is the heterotrophic cycle involving soluble and non-living particulate organic matter, bacteria and microzooplankton. This neglected system which runs parallel to the main food chain may not merely alter the energy flow but could have important feed-back effects for stability. We guess that it is 10—20% of the main system and since this is far below the discrimination achievable in the data, we are forced to ignore it. In terms of this basic structure of models, the state of the art has not really advanced so very far beyond the work of Gordon Riley, 25 years ago. The question is whether the variability is unavailable to modelling and so, in turn, whether these simple models are adequate given this noise in the data.

Any model requires some picture of the physical environment as a basic input. In nearly all models this picture is excessively simple, with all the complications occurring in the biological formulations. O'Brien and Wroblewski [6] showed how a biological model could be built into a two-dimensional picture of the vertical and horizontal circulation of water in a section at right-angles to the Florida coast. Such combinations of physical and biological modelling need technical ingenuity but illustrate the potential for the incorporation of spatial variations. However, they still give smooth changes in parameters and still leave the problem of the "roughness" observed when continuous measurements are made across any area. In relation to this problem Platt [12] (see also Platt, 1972) describes very interesting results using fluorometry which provides a continuous record of plant pigments. From such continuous records he deduced the same spectral structure for chlorophyll variations for the scale 10 m to 1 km as is predicted from theories of turbulent lateral diffusion. Thus some of the small-scale patchiness in data may be explainable in terms of purely physical processes of diffusion. As Kierstad and Slobodkin (1953) pointed out in relation to red tides, for a sufficiently large patch the rate of plant growth within the patch may counteract the dispersive action of turbulence, so that the patch will persist. The critical parameter is the scaling factor used for lateral turbulence since this, together with plant growth rate, will determine the critical patch size. It is known that the diffusion coefficient, K, used to describe turbulence is related to the scale of turbulence. The estimates of this relation have been derived mainly from studies of the spread of dye patches. Okubo (1962) brought together these results and compared them with theoretical descriptions. I have added to his figure the relation indicating critical patch size (l) as a function of diff sion coefficient (K) for a plant growth rate of one division per day (Fig. 10.1). This gives an intercept with observed diffusion coefficients at about 10 km. For lower division rates the critical patch size would be larger. This comparison would imply that patchiness in open waters which had a biological origin, rather than resulting from physical processes, is likely to occur on scales greater than 10 km. There is, however, a second feature of Fig. 10.1: the small angle of intercept between the line for critical patch size and the observed data (or the theoretical descriptions). Thus small reductions in turbulent diffusion imposed by coastal topography or by hydrographic boundaries could decrease the critical patch size considerably.

There are three consequences of these interactions between physical diffusion and biological growth parameters. Firstly, the problem of the stability of populations is usually studied in terms of the nature of biological "functional" responses although the effects of spatial heterogeneity have been recognised as a possible alternative or additional mechanism. As a preliminary hypothesis it could be suggested that turbulence can eliminate small-

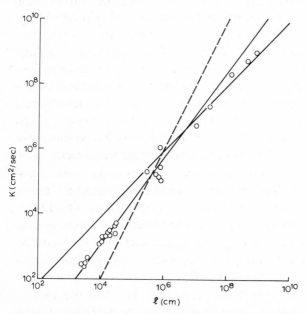

Fig. 10.1. Observed values of diffusion coefficient K, as a function of scale l, with two theoretical relations (from Okubo, 1962). The dashed line indicates the critical value L^* (as a function of K) which determines whether a perturbation will increase or decrease (see text).

scale perturbations (or "patches") but will not be an adequate mechanism to control the larger-scale perturbations such as the spring outburst in the open sea. Secondly, these ideas suggest that models including scale factors could produce, instead of single values for chlorophyll or zooplankton at any instant in time and position in space, spectral compositions over a range of linear dimensions with peaks predicted at certain scales which would not be expected in comparable data on purely physical parameters such as salinity. Thirdly, the numerical values indicated in Fig. 10.1 give some idea of the size of area which would need to be sampled. If the 10-km value were used as a first guess, then a 100-km^2 area would need to be sampled with sample spacing not greater than 1 km.

Data collected on this scale would describe but not completely elucidate the biological importance of patchiness. It is here that large-scale experimental ecosystems can play an important complementary role. Such experimental containers are still very small compared with the scales discussed above. Whether we can or cannot maintain reasonably natural mixed populations in such containers will provide a test for whether patchiness is merely a consequence of imposed physical effect or is an essential part of the life cycle of plants, herbivores and carnivores. The prediction, on the basis of the reasoning given here, is that, although populations within a container should fluctuate more than those outside due to the removal of lateral mixing, these fluctuations should not result in the elimination of major components of the ecosystem.

These comments have indicated how a more realistic picture of the physical processes in the sea could improve the descriptions provided by models. At the other extreme, in building models, there is the problem of the adequate definition of the means by which carnivores remove from the system the organic matter introduced through photosynthesis. In most, if not all, models of the basic productive processes in pelagic systems, carnivores occur as a constant or an independently varying function. They are used as a sink for sources of energy created elsewhere. Yet in many other ecosystems carnivores are credited with a major, if not the ultimate, role in determining the structure of lower trophic levels and the stability of the whole ecosystem. Paine [13] (see also Paine, 1969a,b) described the importance of carnivores in intertidal communities with conclusions similar to those often derived for terrestrial ecosystems. I have suggested (Steele, 1972) that for pelagic marine life the effect of carnivores on community stability may not be so predominant, thus partially justifying their neglect in our models. This neglect, however, arises more from ignorance than from substantial observation or experiment. Even if this neglect has some justification in the study of plant—herbivore interactions, there is still the inverse question of the effect of variations in herbivore density on the populations of particular carnivores, especially those which we harvest as food. Cushing [14] (see also Cushing and Harris, 1973) discussed this problem in relation to fish larvae where it has considerable practical importance for recruitment of stocks to the commercial fisheries. Cushing referred to theories (Jones, 1973) which show that larvae which feed on copepods appear to need copepod densities well above the average normally found, and to need these densities over periods of weeks after the end of their yolk-sac stage. If their chance of finding an adequate food concentration each day was random then mortality rates could be near 80% per day rather than the observed rates of about 10% per day. One possible answer to this problem arises from the earlier discussion of patchiness. If some of the larvae, at the start of feeding, find plankton patches with high enough food densities then, for sufficiently large patches, the rate of loss of larvae from the patch could correspond to the required mortality rate. Cushing and Tungate (1963) have shown that there are large plankton patches which exist for sufficiently long periods of time for this to be a possibility. In other words, the combined effects of turbulent diffusion and plankton patchiness could provide a larval mortality rate due to inadequate food outside the patch, as required by Jones' theory. Thus patchiness, which is probably linked to physical phenomena, could be of importance for the population stability of certain carnivore populations.

When one considers the full life cycle of pelagic fish carnivores such as herring, we are again faced with the problem of differing time and length scales since models which would attempt to link changes in fish stock density with their environment must run for a relatively long sequence of years and consider larger regions within which the fish stock lives. Thus an increasing number of trophic levels in a single conceptual framework may require, technically, the coupling of models each of which operates within different time and length scales. This problem can be seen in the work of MacKinnon and Mann [15], where the full ecological data available for bays in Eastern Canada involves not only plankton but also lobster and plaice populations. The same general problem arises in linking river flow, plant and herbivore production in the Fraser River plume, with the effects of variations in herbivore species on density of juvenile fish (primarily young salmonids) which arrive in large numbers annually to feed on this zooplankton (LeBrasseur et al., 1969; Parsons et al., 1969a,b). Preliminary results from De Lange Boom et al. [16] of a model of the effects or river flow on plankton production suggested that grazing was not an important factor governing the distribution of chlorophyll a in the estuary. Presumably, in the short term, river flow and mixing dominate the chlorophyll distributions. But in the long term it has been shown that variations in zooplankton may affect salmonid growth. Some introduction of other factors including, as De Lange Boom et al. suggested, species composition will be required to link events on the different time scales.

This review suggests that there are technical problems concerning the construction of computer models simulating events in the marine biological environment. Some of these arise from the need to introduce more realistic pictures of the physical environment involving not only spatial variations in one, two or three dimensions, but also the spectral composition of distributions which may provide the best comparison with observations and give insight into the phenomena of spatial heterogeneity. Other technical problems arise from the coupling of models having different time scales particularly those describing the longer-term changes which can occur in deeper water or with carnivorous fish populations.

Beyond these technical questions, however, the main difficulties that emerge are concerned with the data rather than with the models. Theory can indicate what kind of laboratory-based information is required, for example on plankton respiration; it can formulate the scale and intensity of field observations; it can also show the role of large experimental ecosystems and the hypotheses which these would test. The recent proliferation of theoretical studies suggests that these experimental and field programmes are becoming the limiting factor in the general development.

PAPERS SUBMITTED TO THE CONFERENCE

[1] Ulanowicz, R.E., Flemer, D.A., Heinle, D.R. and Mobley, C.D. The systematic modelling of mesohaline planktonic systems.
[2] Schneider, E.D. and Phelps, D.K. Biologic modeling experiments as applied to stressed marine systems.
[3] Von Bröckel, K., Smetacek, V. and Zeitzschel, B. Ecological studies of the plankton in Kiel Bight — field studies and large-scale in situ experiments.
[4] Mountford, K. Predictive equations for gross phytoplankton photosynthesis in two temperate estuaries.
[5] Pichot, G. Etablissement d'un premier model máthematique decrivant l'évolution de l'écosystème "Spuikom".
[6] O'Brien, J.J. and Wroblewski, J.S. A simulation of the mesoscale distribution of the lower marine trophic levels off West Florida.
[7] O'Connor, D.J., Di Toro, D.M. and Mancini, J.L. Mathematical model of phytoplankton population dynamics in the Sacramento—San Joaquin Bay Delta: preliminary results and applications.
[8] Buckingham, S.L. A plankton production model for the Western Gulf of St. Lawrence.
[9] Jansson, B.-O. Preliminary analog simulations of the Baltic ecosystem.
[10] Eppley, R.W. Phytoplankton physiology: a guide to the literature for modelers.
[11] Fee, E.J. Modelling primary production: an empirical approach.
[12] Platt, T. I. The concept of energy efficiency in primary production. II. The influence of physical processes on the spatial structure of phytoplankton populations.
[13] Paine, R.T. On the structure and organization of a rocky intertidal community.
[14] Cushing, D.H. A model of density-dependent processes in plaice larvae.
[15] MacKinnon, J.C. and Mann, K.H. Ecological studies of Eastern Canadian marine inlets: the role of systems models.

[16] De Lange Boom, B., Takahashi, M., Le Blond, P.H. and Parsons, T.R. A mathematical model of phytoplankton production in the Fraser River plume.

REFERENCES

Cushing, D.H. and Harris, J.G.K., 1973. Stock and recruitment and the problem of density dependence. *Rapp. P.-V. Réun. Cons. Perm. Int. Explor. Mer*, 164.
Cushing, D.H. and Tungate, D.S., 1963. Studies on a *Calanus* patch. I. The identification of a *Calanus* patch. *J. Mar. Biol. Assoc. U.K.*, 43: 327.
Jansson, B.-O., 1972. Ecosystems approach to the Baltic problems. *Ecol. Res. Comm. Swed. Nat. Sci. Res. Counc., Bull.*, 16. (Stockholm).
Jones, R., 1973. Stock and recruitment with special reference to cod and haddock. *Rapp. P.-V. Réun. Cons. Perm. Int. Explor. Mer.* 164.
Kierstad, H. and Slobodkin, L.B., 1953. The size of water masses containing plankton blooms, *J. Mar. Res.*, 12: 141.
LeBrasseur, R.J., Barraclough, W.E., Kennedy, O.D. and Parsons, T.R., 1969. Production studies in the Strait of Georgia. Part III. Observations on the food of larval and juvenile fish in the Fraser River plume, February to May, 1967. *J. Exp. Mar. Biol. Ecol.*, 3: 51—61.
Margalef, R., 1973. Nato Science Committee Conference on Modelling of Marine Systems, Ofir, Portugal, 1973.
Okubo, A., 1962. Horizontal diffusion from an instantaneous point-source due to oceanic turbulence. *Chesapeake Bay Inst. Tech. Rep.*, 32, Ref. 62—22.
Paine, R.T., 1969a. A note on trophic complexity and community stability. *Am. Nat.*, 103: 91—93.
Paine, R.T., 1969b. The *Pisaster—Tegula* interaction: prey patches, predator food preference, and intertidal community structure. *Ecology*, 50: 950—961.
Parsons, T.R., Stephens, K. and LeBrasseur, R.J., 1969a. Production studies in the Strait of Georgia. Part I. Primary production under the Fraser River plume, February to May, 1967. *J. Exp. Mar. Biol. Ecol.*, 3: 27—38.
Parsons, T.R., LeBrasseur, R.J., Fulton, J.D. and Kennedy, O.D., 1969b. Production studies in the Strait of Georgia. Part II. Secondary production under the Fraser River plume, February to May, 1967. *J. Exp. Mar. Biol. Ecol.*, 3: 39—50.
Platt, T., 1972. The feasibility of mapping the chlorophyll distribution in the Gulf of St. Lawrence. *Fish Res. Board Can., Tech. Rep.*, 332: 8 pp. (20 fig.).
Riley, G.A., Stommel, H. and Bumpus, D.F., 1949. Quantitative ecology of the plankton of the western North Atlantic. *Bull. Bingham Oceanogr. Coll., Yale*, 12 (3): 1—169.
Steele, J.H., 1972. Factors controlling marine ecosystems. In: *The Changing Chemistry of the Oceans. Nobel Symposium 20.* Almquist and Wiksell, Stockholm, pp. 209—221.
Steele, J.H., 1974. *The Structure of Marine Ecosystems.* Harvard University Press, Cambridge, Mass., 125 pp.
Strickland, J.D.H., 1972. Research on the marine planktonic food web at the Institute of Marine Resources: A review of the past seven years of work. *Oceanogr. Mar. Biol. Ann. Rev.*, 10: 349—414.

CHAPTER 11

MODELLING OF LAKES AS PHYSICO-BIOCHEMICAL SYSTEMS — PRESENT LIMITATIONS AND NEEDS*

C.H. Mortimer

11.1. INTRODUCTION

For two reasons, comments on the present status of limnological modelling are not out of place in a volume devoted to marine systems. First, an
essential marine/fresh-water unity appears when we examine the mechanisms
of interchange of mechanical and thermal energy and of biological production, although the scales and boundary conditions may greatly differ. Second, in the present infancy of whole ecosystem modelling, more effort and
funds are probably being put into lake ecosystem and management models at
the present time — if the Great Lakes are included — than into their marine
counterparts. And further, while the lake models merit study in their own
right, they can frequently serve as convenient test-beds for methods and
verification procedures applicable in marine contexts. Such marine/fresh-
water interchange is likely to remain a two-way one, because in some instances simpler boundary conditions make the marine case more tractable,
while in others the lake model is much easier to verify experimentally.

Introduction of the word "verification" in the very first paragraph ex-
poses the point of view, or bias perhaps, which permeates this essay, namely
that any model remains an intellectual plaything, of limited impact on main-
stream development in aquatic ecology, unless it can be tested and verified
by experiment, or by field observation, or both. A combined, two-pronged
approach — modelling coupled with direct study of the system or its components — holds the greatest promise of success. This rather obvious conclusion
needs frequent re-emphasis in view of the present predilection of environment decision makers and research funding bodies (local and national) to put
considerable faith, hope, and charity (financial support) into large modelling
programs, without comparable support for other approaches to the basic
understanding of lake physics and biology. By the time that an ambitious

---

* Contribution no. 102 of Center for Great Lakes Study, The University of Wisconsin,
Milwaukee, Wisc. (U.S.A.).

but poorly underpinned modelling scheme is seen to have feet of clay — arising from inadequate quality and quantity of input data, or from failure to encompass the relevant time scales or to properly understand the principal controlling mechanisms — large resources will have been dissipated and future support for modelling efforts will be more difficult to obtain. We must therefore find a balance even though, as is usually the case, we can only dimly perceive where that balance should be struck. Experience also points to the need for professional systems modellers and limnologists both to develop the equivalent of the Hippocratic Oath in selling their expertise to environment managers. Modellers should display, at the outset, the confidence limits to be expected for model predictions; and limnologists should identify those key mechanisms which are least understood and also draw attention to the danger of mass collection and "systems analysis" of survey data, if this leads to the neglect of research on fundamental processes.

The desired predictive and control capability is likely to come most quickly through the *combined* devotion of the best talents of "natural historians" (taxonomists, chemists, hydrodynamicists, field investigators) and "modellers" (laboratory experimentalists and numerical analysts). This essay is therefore a plea, not only for the desirability, but also for the *necessity* of that collaboration and for continuing support of both arms of this dual attack on basic problems. The need for balance and for putting the horse (ideas, concepts, and direct lake studies) before the cart (analytical methodology) can, I believe, be illustrated by two brief but revealing, already published examples, one from lake physics, the other from lake biology. But first, something should be said about what constitutes a "model" in this context.

Models take many forms, among which:

(1) *Conceptual models* can be equated with the general logic of the scientific method — development of testable and refutable hypotheses. Systems-thinking has long been a feature of the way in which pioneer limnologists approached lakes. E.A. Birge, impressed early by Forbes' concept of the aquatic habitat as a "microcosm", wrote in his 89th year (1940, pp. 46—47): "Limnology in Wisconsin is the story of the small lakes — of the *lakelet*. For in the lakelet are found all the differentials which make limnology a distinct science... It is the [infinite riches in a] little room of the lakelet which [have] so accentuated both its beauty and its science that even our dullness does not wholly miss them". Also, Forel (1892) wrote "Un lac se rapproche par ses proportions d'un laboratoire où le naturaliste peut répéter à volonté ses recherches, instituer des expériences, interroger la nature au lieu de se borner à en écouter les leçons". In other words, he proposes to ask the lake directly how *it* solves the differential equations.

(2) *Mechanical models*, one or more mechanisms can be qualitatively and

sometimes quantitatively studied in the laboratory (e.g. Von Arx, 1952; Mortimer, 1954; Fultz, 1961).

(3) In *"back-fitted"* models, the parameters and coefficients are adjusted to fit observed behavior of a particular aquatic system. For this reason they give useful predictions for that system as long as it is not strongly perturbed, but may perform poorly in other situations and may not contribute substantially to universal understanding.

(4) *"Black box"* or *extrapolation models*, sometimes developed from type 3 models, relate whole-system ("black box") responses to changes in input. They have the merit of taking complex, non-linear interactions in their stride; but, to exploit this capability, they require an extensive data base of high quality and often of high resolution in time and space.

(5) In *analytical models*, more or less "realistic" assumptions permit actual lake processes or actual basin shapes and boundary conditions to be simulated by mathematical expressions, capable of further manipulation in their given "continuous" form.

(6) *"Discretized"* simulation models are characterized by transformation of the governing differential equations into difference equations and by representation of the system boundaries by rectilinear or other simple geometric approximations, so that any basin shape and boundary conditions can be accommodated in computational schemes, simple in principle, but often vast in execution.

Each of the above six classes of models possess particular advantages and limitations; and one class can often complement another. Models of class 5 are the more general, while models of class 6 provide computer-simulated applications to particular basins and systems. Incidentally, analytical models of class 5 can often serve as essential checks on the development of methodology in class 6.

Rather than attempting a general review of limnological models or a detailed description of representative cases, this essay makes its point through a closer look, in the following sections, at one physical and one biological illustration of lake characteristics, of the kind which models must become capable of adequately reproducing before useful ecosystem prediction and control is possible.

11.2. THE THERMOCLINE AND ITS INFLUENCE ON BIOLOGICAL PRODUCTION

A comprehensive review of this question would, of course, require treatises on physical and chemical limnology; and here we can only list some main headings, pointing where appropriate to the implications for the present state of the modelling art.

Governing *primary production* are the radiation field and nutrient supply. Adequate models exist for the former in homogeneous water and are under development for non-homogeneous cases (Fee, 1973). As noted below, turbulence introduces a stochastic element which determines the *probability* of radiation exposure for each cell, so that primary production rates are strongly dependent on the depths of the euphotic zone and of the "upper mixed layer", as originally pointed out by Gran and incorporated in the more comprehensive model of Riley et al. (1949).

Models of nutrient uptake must incorporate Liebig's limiting nutrient concept in original or modified form, but often such models cannot be sharply defined, either because the nutritional requirements (inorganic and organic) of the particular organisms are not precisely known, or because the chemistry of the element in question is poorly understood in extreme aqueous dilution (for example the silica cycle or the question of which forms of phosphorus are "biologically active"), or because the only available uptake coefficients, based on laboratory culture experiments, are not representative of conditions in a highly turbulent natural environment. These are all research areas in which much more fundamental work is needed before model precision can be improved.

The role of turbulence has already been touched upon. In fact, turbulence regulates to a great extent the properties of most natural environments. But, because of the indeterminacy of motion of individual fluid elements, turbulence can only be treated in terms of statistical models: Fickian and non-Fickian (Richardson and Stommel, 1948) diffusion; cascading of turbulent energy across the eddy spectrum (Kolmogoroff, 1941); the fundamental distinction between "stirring" and "mixing" (Eckart, 1948). In lakes and seas, many properties of the physical environment, important for living organisms — transport and dispersal, nutrient availability, duration of suspension in the euphotic zone, distribution and seasonal cycling of carbon, oxygen, and nutrients — depend very strongly on the properties of flow in *stratified* fluids, i.e. on the interactions between two agents: (1) mechanical turbulent energy, delivered to the water surface by the wind; and (2) the buoyancy field created by the net vertical flux of heat, resulting from the combined effects of radiation balance, exchanges of sensible heat, and the consequences of evaporation. The balance between these two agents, both acting across the air—water interface, determines: the depths to which the wind can mix a given heat income; the formation and seasonal depth progression of the thermocline; the extent of the upper mixed layer (epilimnion) in direct contact with the atmosphere; and the extent and duration of the temporary isolation of hypolimnetic (sub-thermocline) water masses.

The formation and destruction of the thermocline are events of prime importance for the biological regime of lakes and seas; and several one-dimensional models are available to reproduce these events and to provide insight into the mechanisms concerned. Such models treat: interactions between buoyancy fluxes and the fluxes of mechanical turbulent energy (application of the Monin-Obukhov length scale by Sundaram et al., 1969); thermocline formation (Munk and Anderson, 1948; Krauss and Turner, 1967); and entrainment of sub-thermocline layers leading to eventual destruction of thermoclines (Kato and Phillips, 1969). But the simplifying assumptions underlying such models — one-dimensionally, nearly steady states of flow — weaken their application to closed basins, or nearshore waters, in which the boundaries impose a two- or three-dimensional pattern on motion, and in which meteorological forcing, particularly storms, result in flow states which are far from steady.

The inadequacies of one-dimensional thermocline models may be illustrated by the previously published (Mortimer, 1954, 1974) qualitative description of the probable stages in development of a commonly observed pattern of isotherm distribution (Fig. 11.1D) after sustained application of wind stress to the surface of a stratified lake.

The extent of the upper mixed layer (shown unshaded) depends on the depth to which the (mechanical) turbulent energy of the wind in previous storms has been able to transport the previous heat income, entering at the surface, against the buoyancy forces associated with density gradients which that heat income produced. The depth of the resulting mixed layer, in relation to that of the euphotic zone, exerts the primary hydrodynamic control on algal photosynthesis.

At the bottom of the turbulent mixed layer, a relatively steep density gradient (the thermocline) offers strong impedence to mixing between that layer and the lower (shaded) hypolimnion. As this normally lies below the euphotic zone, it is not usually the seat of organic production, but rather the seat of organic decomposition, oxygen uptake, and nutrient release. But, unless the thermocline becomes "leaky", during storms and through mechanisms to be described later, exchange between the hypolimnion and the upper illuminated, oxygenated layers remains strongly suppressed by the steep density gradient in the thermocline.

Fig. 11.1 illustrates the essential two-dimensional nature of the modelling problem. Because the basin has boundaries, application of the wind stress forces the upper and lower layers to move in opposite directions, producing a small tilt of the water surface, a larger tilt (of opposite sign) of the thermocline, and an epilimnetic wedge with a circulation pattern similar to that illustrated in sections B and C — a surface current in the downwind direction, balanced by an upwind return current just above the thermocline. If, relative to the thermocline density gradient, the shearing action of that return current rises above a critical value, the nature of the flow suddenly

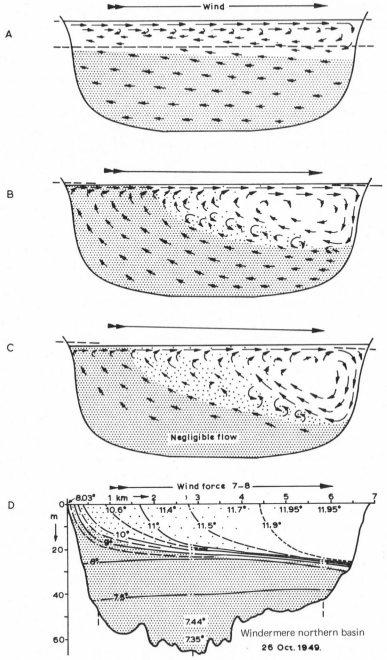

Fig. 11.1. Two-dimensional representation of the effects of wind on a stratified lake. A—C. Stages in the development of the epilimnetic wedge and of shear instability at the downwind end of the thermocline. D. Isotherm distribution in Windermere, northern basin, after about twelve hours of relatively steady wind. The layer initially below the thermocline is shown stippled, and relative speeds and direction of flow are roughly indicated by arrows. (From Mortimer, 1961.)

changes from relatively smooth flow, in which turbulence is suppressed, to a condition of "shear instability" characterized by the growth of small internal wave disturbances into large vortices, illustrated by curved arrows in Fig. 11.1 and vividly reproduced in Thorpe's (1971) experiments. These confirmed that shear instability sets in when a non-dimensional parameter, the Richardson number, falls below a critical value.

The steepening of the thermocline at the downwind end of the basin (Fig. 11.1D) can be interpreted as a result of localized shear instability, which acts like a carpenter's plane to erode the thermocline and to entrain water from sub-thermocline layers. The resultant "shavings" — large unstable vortices disintegrating into packets of relatively intense turbulence — then drift upwind with the mean flow of the "return current", to produce the fan-shaped distribution of the isotherms at the upwind end; and the whole process represents a temporary form of thermocline "leakage", i.e. a transfer of sub-thermocline water (containing nutrients, perhaps) into the upper layers. Another form of leakage is the upwelling (also temporary) of hypolimnetic water to the surface at extreme upwind end of the basin (at the left in Fig. 11.1B,C and D).

When the wind stress is removed, a redistribution of the displaced and newly formed layers takes place, involving return flows and internal seiche oscillations, which will not be described here in detail. (If the amplitudes of the internal seiches are great enough, shear instability may also set in at the seiche nodes, representing yet another form of temporary thermocline leakage.) The important point for the would-be modeller to note is that the disposition of the isotherms *after* the disturbance has past, i.e., the shape and depth of the thermocline on any given vertical along the lake axis, depends principally on happenings elsewhere, i.e. on the following *combined* influences: (1) whether or not shear instability occurred, and on its location and extent, principally at the downwind end of the basin; (2) on the subsequent advective redistribution of the mixed water, newly created at the downwind end; (3) on upwelling, if present, at the upwind end; and (4) on the internal seiches which the storm-displaced layers also set in motion. The combined effect of these processes and series of events, in the later part of the lake's stratified season, causes the thermocline to descend, not regularly, but in steps after each storm (Mortimer, 1961).

The importance, for lake processes, of storms and other severe meteorological perturbations has generally not been sufficiently emphasized. A few hours of storm can achieve more in the redistribution of water masses — leakage of hypolimnetic, nutrient-rich water into surface layers, for example — than several weeks of calmer weather. Two important conclusions emerge from the scenario sketched in foregoing paragraphs. One, for the physical limnologist, is that a great deal remains to be observed about the flow pattern in lakes — particularly including measurements of the magnitude of the Richardson number — before these patterns can be modelled with convincing realism. Second, the system analyst must adjust his models to the conclusion that patterns extending over large (whole-basin) space and (seasonal) time scales are often determined by short-term, localized, strong perturbations (e.g. local shear instability) commonly characterized by nonlinearity and by much smaller space and time scales. This has profound implications, not only for modelling, but also for design of verification experiments.

However, now that the governing principles and the complexities of lake physics are beginning to be recognized, there is hope that the design of appropriate *hydrodynamic* models will not be long delayed; and indeed some have already scored predictive successes (Heaps and Ramsbottom, 1966; Rao and Murty, 1970; Simons, 1971; and Platzman, 1972), opening up avenues of knowledge not accessible through field observations of the "classical" type alone. On the other hand, modelling of whole-basin biological (*ecosystem*) events and processes is as yet a long way from achieving a comparable, generally applicable, predictive power. Only in a few relatively simple situations are the biological or managerial models beginning to demonstrate the hoped-for (although often promised) capability of going beyond classical, Forelian limnological conceptualizations based on *direct* study of a lake's own responses, pursued in sufficient detail over space and time. To suggest why this is so, we now turn to our second example.

11.3. BIOLOGICAL ACTORS ON THE PHYSICAL STAGE

A model capable of providing useful predictions for lake managers must also be able to match component models of "relevant" physical processes with other component models which treat the relevant activities of organisms. Relevance is here defined in terms of the need of the user. For example, the components considered in a fisheries model would be different from those treated in an "algal proliferation" model. The model designer, or more probably the design team, must be in a position to do the following: judge which of the lake's responses are relevant for the user; recognize and preferably identify precisely all the organisms involved in those responses; describe in quantitative (rate) terms the nutrient requirements, growth responses, and interactions of organisms (at all relevant trophic levels) and the time-history of supply rates from all nutrient sources, under conditions of natural turbulence; circumvent the difficulties posed by the matching requirements of the physical and biological model components.

A common source of difficulty will be mis-match of time and (to a lesser extent) space scales. For example, whole-basin circulation models are now commonly run with time steps of order one minute (for reasons of computational stability); while models of thermocline instability, for example, may require even shorter time steps. Biological production, on the other hand, can be adequately described using much longer time steps, as units of seasonal time scales; although as pointed out in the next paragraph, the course of biological production may often be *determined* by switching operations, individually occupying only very short intervals of time. This crucial point merits some elaboration.

It is often characteristic of causality in biological systems that they tend to follow a particular path for a relatively long time span, but that the *selection* of that path is an event (as in rail travel) which occupies only a brief switching interval. Limnological examples are: the environmentally controlled switchover from parthenogenesis to sexual reproduction in *Cladocera*; encystment and hatching; onset of epidemic diseases; sudden changes in the feeding selectivity of secondary producers (e.g. zooplankton); population explosions which often follow introduction of exotic species (e.g. the lamprey into the upper Great Lakes); and the sudden "unpredictable" onset of fungal parasitism of algae, described below. Extending the railroad analogy further, progress along the selected path is controlled by fuel (energy and nutrient supply) and track gradients (environmental stresses and boundaries), all acting relatively slowly; but the switch — often determined by quite a different set of factors — requires little time and little energy to complete its operation. An adequate model must, therefore, accommodate these widely divergent scales and processes, at least in a stochastic if not in a fully deterministic manner.

Starting logically at the level of energy conversion and primary production in the world of micro-organisms, the limnological modeller will also have to take the following biological subject-areas within his purview: definition of essential inorganic nutrient and organic micro-nutrient requirements for each relevant organism, determined in axenic laboratory cultures and extrapolated to field dilutions*; physiological peculiarities, particularly those which influence cell nutrient intake and growth interactions. Examples are: turbulence-regulated diffusion gradients at the cell surface (Munk and Riley, 1952; Hutchinson, 1967, Ch. 20); luxury uptake of phosphorus (Mackereth, 1953);

---

* Lund (1959) writes: "We know relatively little about chemical factors and I believe that the greatest need for the future is more concentration on chemistry. Two of our difficulties are our relative ignorance about certain trace elements and almost all the organic compounds... All our culture work, which involves chemistry, will lose much of its value if it cannot be related to the varying chemistry of the natural environment. Indeed it is extraordinarily difficult to devise cultural procedures which will give answers which, from observation, we have not good reason to suspect are inapplicable to natural conditions."

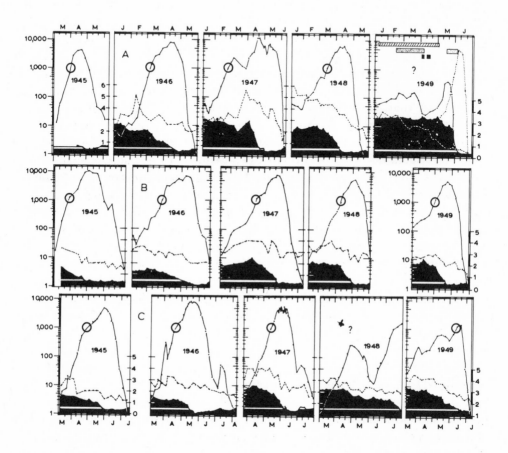

Fig. 11.2. Spring growth of *Asterionella formosa* in the upper 5 m of 3 lakes: A, Esthwaite Water, B, Windermere, southern basin, and C, Windermere, northern basin, with respective maximum depths of 15, 33, and 65 m. Cell numbers per ml are shown (unbroken line) on a logarithmic scale. Circles indicate when the population reached 1000 cells/ml. Concentrations (mg/l) of dissolved silicate (as SiO_2, black area) and nitrate (as $N \times 10$, broken line) are shown on a linear scale. Horizontal white lines within the black areas, indicate the critical concentration of 0.5 mg/l SiO_2. Indications in the top right-hand diagram (Esthwaite Water, 1949) are: hatched area, abundance of the blue-green alga *Oscillatoria*; dotted areas, severe parasitization of *Asterionella* by fungi; black rectangles, floods; dotted line, cell numbers of other diatoms, *Fragilaria* and *Tabellaria*. Question marks shown when the "normal" spring population maximum did not occur. (From Mortimer, 1959, based on figures in Lund, 1950, consult for further particulars.)

inter-specific cooperation (algal—bacterial nutritional relationships, growth-promoting extrametabolites (Lucas, 1958) joint roles of fungi and bacteria in the breakdown of organic matter); inter-specific competitions (e.g. for nutrients) or antagonisms (e.g. production of inhibitory extrametabolites, Lefèvre et al., 1952). Also, species succession is a common if not universal characteristic of seasonal cycles and of man-made perturbations in lakes. To simulate these adequately, a succession of model states, with transitions, must also be envisaged.

Again to illustrate some of these points we now look at a particular example in detail, in this case Fig. 11.2, which illustrates the controlling influences of limiting nutrient concentration and other factors on the spring outburst of diatom growth in three neighbouring lakes in the English Lake District (Lund, 1950): These lakes — Esthwaite Water, Windermere south and north basins — possess not greatly dissimilar chemical characteristics, but differ in maximum depth, 15, 33, and 65 m, respectively. Lund's observations, since continued over many more years than shown in the figure, and demonstrating the great value of long-term study of a lake, disclose a relatively simple situation in terms of limiting nutrient control, but with occasional profound complications, which a simple nutrient model would have been incapable of predicting.

As pointed out in an earlier review of Fig. 11.2 (Mortimer, 1959), the progress of diatom growth in the spring of each year is subject to four types of control. Only the first two of these are easy to model.

(1) *Physical control.* In each basin and year, *Asterionella* growth (shown on a logarithmic scale in the figure) commenced in early spring; and most of the growth took place before stratification set in, i.e., while the water column was fully mixed from top to bottom. The depth of the water column was least (max. depth 15 m) in basin A and greatest (max. depth 65 m) in basin C. Because the cells were distributed by turbulent currents throughout all depths, the average duration of light exposure and therefore cell growth rate was inversely correlated with column depth. Therefore in the shallowest basin, a given population level (1000 cells per ml indicated by circles in the figure) was attained some weeks earlier than in the deepest one.

(2) *Limiting nutrient control.* The normal pattern of the "spring increase", illustrated in the figure, was accompanied by a progressive fall in silicate concentration to a "limiting" level of about 0.5 mg/l SiO_2 , corresponding to maximum population peaks between 5000 and 10,000 cells per ml. Analysis of the silica concentration in the cells confirmed that one more cell division of the population at that level would have required more silicate than was available in the water. Further growth was, therefore, impossible; and the population peak was followed by a die-off phase. For this simple example of nutrient control, it would be easy to devise a model, also taking

the column-depth factor into account, to describe what happened in *most* years. But, as Fig. 11.2 shows, there were exceptional years illustrating other types of control.

(3) *Biological control.* The year 1949 in basin A was marked by heavy parasitism of the diatoms by chytrid fungi (over the intervals indicated by dotted rectangles in the figure); and the silicate concentration in the water did not fall to limiting levels until two *other* diatom species had later appeared on the scene (see figure legend).

(4) *Trace nutrient control.* The spring increase in basin C in 1948, although it started out normally, was halted at a population level of a little over 100 cells per ml, a type of behaviour which Lund correlated with the very dry weather of that year. The most probable explanation of this exceptional behaviour was that a trace nutrient, possibly organic and normally supplied by inflowing streams, was in such short supply that the population could not proceed to the normal level of silicate limitation. In 1948 that level was only reached after a June flood (supplying the missing micronutrient?) permitted diatom growth to restart.

The implications of Fig. 11.2 for limnological modelling are far-reaching, because we must expect similarly varied biological responses, although with different details and with other organisms and other nutrients, to be found in other lakes when studied with comparable intensity and continuity. The incidence of parasitism, for example, may require stochastic modifications of the model; and Fig. 11.2 also illustrates the need to direct attention to the importance of climatic variability and soil processes operating in the lake's catchment area. For example, the decrease in nitrate concentration in the figure appears to be related, not to the growth of the planktonic diatoms, but to processes occurring in littoral regions and in the catchment soils.

Interactions involving phosphorus, usually the limiting nutrient in fresh water, must also be clarified before valid modelling is possible. Not only is there doubt about which fractions are measured by analytical procedures, but also about which of these fractions are available for algal growth. For example, Rigler (1966) presents evidence that the commonly used molybdenum blue method overestimates the concentration of inorganic phosphorus in lake water by one or two orders of magnitude. The implications of this for general limnological modelling, as at present practiced, are profound.

Also frequently overlooked is the general observation that the species compositions of undisturbed lakes can remain relatively unchanged for many years, whereas a disturbance — for example even quite a small addition of nutrient materials or man-made wastes — results in relatively sudden and widespread changes in species composition. While it may be possible to construct an ecosystem model for the unperturbed state, that model would obviously be of little use to predict the influence of perturbations, should these go beyond the critical changeover point at which the existing set of actors leaves the stage, to be replaced by a different set. Therefore, a valid perturbation model — the kind usually desired by lake managers for predictive purposes — must not only be capable of reproducing the unperturbed play, but must also predict which of the multitude of hungry actors waiting in the wings will take over the stage after the perturbation.

11.4. CONCLUDING DISCUSSION

It may seem that the foregoing paragraphs have done little more than draw attention to the nature of the complexity of physical and biological interactions in lakes, pessimistically inferring that whole-lake ecosystem modelling still has a long way to go before universally useful, predictive results can emerge. While this inference is at present inescapable, there are other more positive conclusions to be drawn. At the risk of repeating truisms, these conclusions may be summarized as follows.

First, modelling efforts must be solidly backed by *sustained and intensive investigation* of the physics, chemistry, and biology of lakes and seas. After all, modelling is only one component of the tool kit for ecosystem understanding, prediction and control. It is also a tool, albeit a powerful one, which loses its edge if not nourished by an adequate supply of direct observations and guided by the best physical and biological insight, continually tested by verification in nature. Modelling also has an important service function in the verification process — that of optimizing experimental design.

Second, *the two principal objectives of modelling activity at the system level* must not be lost sight of. These are (1) the creation of tools for gaining deeper insight into the workings of the system, and (2) the development of usefully applicable, predictive capability. In general, the limnologist tends to lay more stress on the first objective, while the environmental manager looks hopefully toward the second. Stated with over-dogmatic brevity, the theme of this essay is that any procedure which does not contribute to one or other of these objectives does not constitute a worthwhile "model". Further, the second objective is not yet in sight in spite of some optimistic claims to the contrary, and will not be achieved without substantial progress toward the first. Before the grand synthesis of "limnological systems analysis" can bear universally predictive fruit, a great deal of effort has to be devoted to the selection, design, and testing of relevant component models, to satisfactory matching of the components, and to continuing experimental and field verification.

Third and last, the desired results can only be achieved by a *partnership of effort.* It has been asserted by some enthusiasts (quoted by Regier, 1974) that the development of systems modelling will exert as profound an influence on biological thinking as did the development of statistical analysis. While others will argue that this analogy is overstretched — because systems modelling will have much less bearing than general statistical analysis on the fundamental concepts and properties of the natural world and modes of investigation — the fact that this kind of assertion is being made is a sign of the emergence of a powerful methodological expertise and capability which must be put to work. But, in doing so, it is also necessary to examine what the method can and cannot do. For example, if it can take into account the species changes which are likely to occur, the method can speedily and repeatedly test the sensitivity of a system to perturbations over ranges which include or go beyond those in nature; it can uncover and display patterns inherent in the data, expose data gaps, and optimize data collection and the design of verification experiments; and it can, in general, provide powerful and fast manipulative and predictive capabilities as tools for research and as bases for enlightened management of problems which affect whole systems. But the method cannot create data and, except within a strict and continuous verification framework, cannot distinguish good input data from bad, cannot judge the validity of assumptions, and cannot generate original concepts. In other words, a computer can never substitute for ideas, although it can powerfully assist in their birth.

The successes and failures in ecological modelling and the efforts which continue in this field are leading to less extravagant claims (of the "give me your data and I will build a model" kind) and to more cautious and qualified optimism:

"If models are put in their proper perspective as one of the necessary tools of studies of marine ecosystems, and if these models and their individual components are tested in the field, it should eventually be possible to use a family of systems models as a guide both to an understanding of the dynamics, and to the prediction and management of perturbations of marine coastal areas. Without a systems approach, it is not immediately clear that systems problems can be solved."   Walsh (1972)

We must therefore conclude, from all that has been said, that there is an obvious need for a partnership between the systems analyst and the "classical" limnologist, including the essential data-gatherers, model verifiers, and also the "prima donna" specialists, at which "interdisciplinary" team research administrators seem often to look askance (Walsh, 1972). Rarely, all elements of that partnership are combined in one person (we must train more such), but more often team work will be needed, based — we must hope — on spontaneous recognition of what each member has to offer and on a minimum of bureaucratic coordination. The worthwhile question (cf. Regier, 1974) is, not whether one partner is more important than the other — for there can be no separate, superior priesthood of systems analysts or limnologists — but whether one partner is lagging behind the other.

REFERENCES

Birge, E.A., 1940. Reply to addresses delivered in his honor, published under the title, *Edward A. Birge, Teacher and Scientist*. University of Wisconsin Press, Madison, Wisc., pp. 32—48.
Eckart, C., 1948. An analysis of the stirring and mixing processes in incompressible fluids. *J. Mar. Res.*, 7: 265—275.
Fee, E.J., 1973. Nato Science Committee Conference on Modelling of Marine Systems, Ofir, Portugal, 1973.
Forel, F.A., 1892—1904. *Le Léman; monographie limnologique*. F. Rouge, Lausanne. Vol. I (1892), 539 pp. and map; Vol. II (1895), 651 pp.; Vol. III (1904), 715 pp.
Fultz, D., 1961. Developments in controlled experiments on large-scale geophysical problems. *Adv. Geophys.*, 7: 1—103.
Heaps, N.S. and Ramsbottom, A.E., 1966. Wind effects on the water in a narrow two-layered lake. *Philos. Trans. R. Soc. Lond.*, 259A: 391—430.
Hutchinson, G.E., 1967. *A Treatise on Limnology*, 2. Wiley, New York, N.Y., 1115 pp.
Kato, H. and Phillips, O.M., 1969. On the penetration of a turbulent layer in a stratified fluid. *J. Fluid Mech.*, 37: 643—655.
Kolmogoroff, A.N., 1941. The local structure of turbulence in compressible viscous fluid for very large Reynolds Number. *Dokl. Akad. Nauk S.S.S.R.*, 30: 301.
Krauss, E.B. and Turner, J.S., 1967. A one-dimensional model of the seasonal thermocline. *Tellus*, 19: 98—105.
Lefèvre, M., Jakob, H. and Nisbet, M., 1952. Autoantagonisme et heteroantagonisme chez les algues d'eau douce. *Ann. Stat. Centr. Hydrobiol. Appl.*, 4: 5—198.
Lucas, C.E., 1958. External metabolites and productivity. *Rapp. P.-V. Réun. Cons. Perm. Int. Explor. Mer.*, 144: 155—158.
Lund, J.W.G., 1950. Studies on *Asterionella formosa* Hass, II. Nutrient depletion and the spring maximum. *J. Ecol.*, 38: 1—35.
Lund, J.W.G., 1959. Investigations on the algae of Lake Windermere. *Adv. Sci., Lond.*, 61: 530—534.
Mackereth, F.J.H., 1953. Phosphorus utilization by *Asterionella formosa* Hass. *J. Exp. Bot.*, 4: 296—313.
Mortimer, C.H., 1954. Models of the flow-pattern in lakes. *Weather*, 9: 177—184.
Mortimer, C.H., 1959. The physical and chemical work of the Freshwater Biological Association 1935—57. *Adv. Sci., Lond.*, 61: 524—530.
Mortimer, C.H., 1961. Motion in thermoclines. *Verh. Int. Ver. Limnol.*, 14 : 79—83.
Mortimer, C.H., 1974. Lake hydrodynamics. In: *50 Years Jubilee Symposium, Int. Assoc. Limnol., Kiel, 1972. Mitt. Int. Ver. Limnol*, 20.
Munk, W.H. and Anderson, E.R., 1948. Notes on a theory of the thermocline. *J. Mar. Res.*, 7: 276—295.
Munk, W.H. and Riley, G.A., 1952. Absorbtion of nutrients by aquatic plants. *J. Mar. Res.*, 11: 215—240.
Platzman, G.W., 1972. Two-dimensional free oscillations in natural basins. *J. Phys. Oceanogr.*, 2: 117—138.
Rao, D.B. and Murty, T.S., 1970. Calculation of the steady state wind-driven circulation in Lake Ontario. *Arch. Meteorol., Geophys., Bioklimatol., Wien*, A19: 195—210.

Regier, H., 1974. Fish ecology and its application. In: *50 Years Jubilee Symposium, Int. Assoc. Limnol., Kiel, 1972. Mitt. Int. Ver. Limnol.*, 20.

Richardson, L.F. and Stommel, H., 1948. Note on eddy diffusion in the sea. *J. Meteorol.*, 5: 238—240.

Rigler, R., 1966. Radiobiological analysis of inorganic phosphorus in lakewater. *Verh. Int. Ver. Limnol.*, 16: 465—470.

Riley, G.A., Stommel, H. and Bumpus, D.F., 1949. Qualitative ecology of the plankton of the western North Atlantic. *Bull. Bingham Oceanogr. Coll., Yale*, 12(3): 169 pp.

Simons, T.J., 1971. Development of numerical models of Lake Ontario. *Proc. Conf. Great Lakes Res., 14th*, pp. 654—669.

Sundaram, T.R., Easterbrook, C.C., Piech, K.R. and Rudinger, G., 1969. *An Investigation of the Physical Effects of Thermal Discharges Into Cayuga Lake*. Cornell Aero. Lab., Buffalo, N.Y., 206 pp. (Rep. No. VT 2616-0-2).

Thorpe, S.A., 1971. Experiments on instability of stratified shear flows: miscible fluids. *J. Fluid Mech.*, 46: 299—319.

Von Arx, W.S., 1952. A laboratory study of wind-driven ocean circulation. *Tellus*, 4: 311—318.

Walsh, J.J., 1972. Implications of a systems approach to oceanography. *Science*, 176: 969—975.

PART III

REPORTS OF THE WORKING GROUPS AND RECOMMENDATIONS
FOR FUTURE WORK

CHAPTER 12

SPATIAL INHOMOGENEITY IN THE OCEANS

J.J. O'Brien (*Group Leader*)
T. Platt (*Rapporteur*)

P. Le Blond	W. Krauss	W. Zahel	J. Steele
R. Margalef	D.H. Cushing	P. Liss	A. Pires
J. Walsh			A. Finza

12.1. INTRODUCTION

Recent theoretical and field studies indicate that the horizontal length-scales relevant to this discussion are in the range 10^2 to 10^4 m.

The important vertical scales are 10^{-1} to 10^2 m. The time scales of interest are from 10^4 to 10^7 sec, covering the range from the generation time of phytoplankton to the lifetime of the largest zooplankton patch ever studied systematically. It was recognized that this range of horizontal length-scales is the least understood in all of physical oceanography. Spatial inhomogeneities on these scales are extremely common in biological oceanography and are probably not uncommon in physical oceanography.

The relevance of the word "model" in this context is restricted to the theoretical, explorative model. By this we mean the systematic study of ideal cases to elucidate the kinds of solutions (analytic, stochastic or numerical) which can exist, and therefore the kinds of spatial structure which can occur for given combinations of the critical parameters. Large and complex simulations designed to reproduce or predict in detail the patchiness conditions in specific areas of the real world are excluded from our consideration as being premature. It is recognized, however, that the results of exploratory, theoretical analysis will be useful in the parameterization of spatial fluctuations in more complex models.

The kinds of models envisaged by us could be studied easily and economically on existing computers.

12.2. RESEARCH AREAS

Areas of research which merit particular emphasis to facilitate progress in the study of spatial inhomogeneity in the ocean include:

(1) A focus of attention of both theoretical and practical physical ocean-

ographers on this poorly understood 10^2 to 10^4 m region of horizontal length-scale.

(2) An improved knowledge of magnitude of the diffusion coefficient (both horizontal and vertical) in a variety of fields.

(3) There is a definite need for exploratory modelling to go on hand-in-hand with field studies of spatial inhomogeneity carried out in a multi-disciplinary manner. This will require the development of continuous-recording (in situ where possible) automatic instruments for the measurement of biological, chemical and physical quantities in sea water. Field studies designed to observe and characterize the biological structure on the larger length-scales will require a considerable effort at least multi-ship, and possibly multi-national.

(4) Studies on the relationships between the wind field and the motions of the upper ocean should be continued and expanded.

(5) A specific question which is of immediate interest, but for which no answer is available is "What is the relationship between the size and lifetime of spatial inhomogeneities?". The relative importance of physical and biological processes in deciding this question should be established.

12.3. IMPLICATIONS FOR MANAGEMENT AND POLICY-MAKING

The implications for management and policy-making of these studies include:

(1) Knowledge of the dynamics of formation and decay of patches in the sea bed applies directly to problems involving the introduction of contaminants into the sea.

(2) Since there appears to be a minimum linear scale of spatial inhomogeneity which is critical for the stability and continued growth of patches of phytoplankton, the results of these studies are of importance in decisions about the size and spacing of, for example, sewage outfalls along the coast.

(3) A knowledge of the statistical properties of the distributions of chemical, physical and biological quantities in the sea is an essential prerequisite of the design of adequate and economical sampling surveys for the monitoring of water quality on the oceanic scale.

(4) Recent models in fisheries biology recognize the importance of spatial structure in the distribution of the food availability to larval fish. The inhomogeneity stabilizes the production system by ensuring that a sufficient proportion of the larvae find an adequate food supply in regions where the average supply is below the level required to prevent starvation.

CHAPTER 13

COASTAL OCEANS

B. Zeitzschel (*Group Leader*)
I.N. McCave (*Rapporteur*)

L. De Conninck   T.S. Hopkins
J.C. MacKinnon   A. Piro
R.T. Paine   J. Dugan
A. Distèche   L. Saldanha

13.1. INTRODUCTION

We take shelf seas to a depth of 200 m, excluding estuaries and enclosed seas, to be the area under consideration. The peculiarities of this zone which render it both important and complex are the facts that it is shallow, intensively used by man, is a region of high productivity, and is adjacent to the land with all that that implies. The boundaries of the region include a possibly deformable bed and coast, the edge of estuarine regions, the air—sea interface and the open ocean boundary through all of which exchanges of energy, inorganic and organic matter occur.

Many classes of model have been developed to deal with particular aspects of this region. These include fisheries, productivity, coastal dynamics, storm surge and tidal models. The *desideratum* is a differential model which is of sufficient generality that it will be able to incorporate wherever possible all these aspects of the coastal ocean, allow interactions between them, and respond accurately to perturbations which go beyond the range of the values used to validate the model. Particular aspects of the total model which are not critically dependent upon other components (e.g. physical models) have shown some degree of success. It is in the interactive models, in particular the ecosystem model which is dependent on many non-linear and behavioural factors, that we find greatest difficulty. There are some classes of biological problem (e.g. the predator problem) which will have to be formulated at the species level and for which community effects will have to be taken into account. The flow of energy and cycling of matter through the system is an essential feature of interactive models involving ecosystems.

We now consider some of the problem areas which require attention before general interactive models can become a reality.

13.2. PROBLEMS AT THE BOUNDARIES

The deformability of the coast is successfully treated in wave-power models of longshore sand transport. However, the movement of material in an on—offshore sense is not well understood. As the seasonal cycle of this movement may affect nearshore benthic communities, it merits early attention. The supply or exchange of material (including biota) at the mouths of streams, estuaries and coastal inlets needs to be understood, particularly in respect to its temporal variability (e.g. the suspended and dissolved loads of rivers at varying time scales including flood run-off conditions). Neither the estuary nor the coastal inlet sites can be isolated from the coastal ocean in models as biotic exchanges at different stages of the life cycle are of major importance.

Benthic boundary layer studies are urgently needed. These must include the effects of fair-weather and storm conditions on water motion and shear stress near the bed, the effect of shear stress on resuspension of the bed, the effect of the water movement and resuspension on benthic communities (and the effect of benthos on resuspension), as well as the kinetics of chemical adsorption—desorption reactions involving suspended material.

Data and models for transfer of mass and momentum through the air—sea interface should be incorporated in the ecosystem model. The stability of the thermocline and associated mixing under normal and storm conditions requires theoretical and multidisciplinary field investigations.

Prediction of the extent of ice cover in coastal oceans is an important field for future modelling activity. Its effects are felt on the dynamics of the circulation, on sea surface phenomena including transfer mechanisms, and in disturbance of benthic and coastal communities. The storage, transport and release of material, together with modification of patterns of salinity, heat and plankton productivity distribution are important aspects of the ice problem. These phenomena together with the effects of ice on shipping and other human activities also require incorporation in models.

13.3. PROBLEMS IN THE INTERIOR

Productivity and nutrient uptake models are reasonably advanced, but chemical and energy transfer coefficients within the food web are poorly known at present. The large-scale exchanges with the open ocean are poorly predicted by existing oceanic models, and the data base for specification of these exchanges does not exist in the majority of areas. The volume and source of upwelling water in coastal oceans are vital factors in the formulation of models involving productivity at all levels. Fisheries models could at one time be considered in relative isolation, but it is now more urgent to consider all levels of the food web in assessing the effects of man's activities. Thus only the ecosystem model will be adequate to deal with evaluation of possible perturbations of the system. Prediction of the distribution and transport of planktonic populations are critically dependent upon the formulation of working physical models of circulation, and the spatial and temporal distribution of diffusion coefficients.

The inhomogeneity of the coastal ocean when considered in its physical, chemical and biological aspects, dictates a wide range of possible spatial and time scales. Of first-order magnitude is the 10—100-km scale of shelf width. Smaller scales are appropriate and must be defined at the beginning of specific investigations. Time scales obviously involve the seasonal cycle, and the difference in generation-times between whales and bacteria will be apparent.

13.4. POLLUTION AND OTHER PERTURBATIONS

For models of coastal oceans to have a useful predictive capacity they must be able to respond accurately to perturbations. Pollution is one such perturbation. The relevant kinetic factors, physical and biological effects associated with particulate inert, radioactive and biologically active materials, dissolved hazardous compounds and thermal pollution, should be assembled and made available to all modellers. Total-ecosystem models need to contain a mechanism for signalling when specified quantities of hazardous substances are exceeded in any component of the system. Such signalling would also indicate, for example, imminent eutrophication or unacceptable effects of fishing activities. A problem which must be recognized is that the rate terms which are applicable to normal processes may be incorrect in the case of drastic perturbation of the system.

13.5. ASPECTS OF MODELLING AND ITS USE

We consider the formal framework provided by an attempt to establish a model to be very valuable in the identification of areas where research is needed, and in its subsequent management through planning and data collection phases. Not only in the specification of the required data base (boundary conditions and validation), but also in the recognition of necessary rate and transfer functions, the logical structure of the model is useful. We do not distinguish between classes of models, only the use to which they are put. Thus a management model is the same scientific model of which questions important to management may be asked. If models are to be asked such questions, it is important that those in decision-making bodies who receive the results should be made aware of the limitations which have entered into the formulation of the model. There will be various generations of models. First-generation types will have to include estimated transfer coefficients and be limited in scope. The danger inherent in the prediction of perturbation effects from such models will be apparent to the modeller but must be made clear to others for whom the model may appear as an all-powerful black box. Succeeding models with more detailed kinetic and physiological factors will hopefully emerge from research effort stimulated by first-generation types.

13.6. AREAS OF RESEARCH

A number of fields of research are identified in this report. Those which we feel to be of greatest importance are listed here.

(1) *The storm problem*: This involves the effects of storms with intense wave activity on near-bed flow structure, on resuspension of sediment, on benthos and on thermocline stability. An opportunistic strategy should be developed to monitor extreme conditions.

(2) *The predator problem*: This concerns the effect of predators on organic communities. Artificial introduction or removal of predators produce effects which need to be made predictable.

(3) *Grouping of organisms*: This involves identification of functional groups with determination of rate factors appropriate to each group. Treatment at species level is likely to lead to models of intractable complexity, and therefore will not be included in first-generation models.

(4) *The chemical kinetics problem*: Many of the chemical models indicate only the equilibrium states of the system. Rates of transition between these states are required for modelling purposes. These involve rates of conversion between species, adsorption/desorption rates and other reaction rates.

(5) *Toxic substances*: The sublethal toxicity levels have to be assessed for numerically and functionally important species.

(6) *The recycling problem*: The rate terms associated with cycling of dissolved and particulate material and energy flow through a given biological system are required.

(7) *Diffusion coefficients*: Many problems of plankton and other suspended material distribution require knowledge of spatial and temporal values of diffusion coefficients. Care should be taken to ensure that a part of the research effort in physical oceanography is devoted to provision of these data.

(8) *Boundary exchange*: Investigations in the rate of exchange between coastal waters and adjacent water bodies and in general at interfaces are

required. The cost-effectiveness of alternative theoretical or measurement approaches to meet this deficiency should be investigated.

(9) *Circulation patterns*: The dynamics describing the time-dependent velocity field in each locality are essential information in the formulation of an ecosystem model. Included in this category would be water mass origin and structure, tidal and coastal currents.

(10) *Spatial and temporal heterogeneity*: Model realism requires that physically or biologically induced inhomogeneities be studied and incorporated in ecosystem models.

(11) *The interfacing problem*: Many models of various aspects of the total system already exist. A major theoretical problem which must be tackled immediately is the interfacing of, e.g. biological with physical and chemical models.

13.7. CONCLUSIONS

We find the value of the modelling approach for the coastal oceans to be considerable. We suggest that the approach be fostered by all appropriate means.

We would stress the importance of the interdisciplinary nature of modelling, and suggest that it has an important role to play in the education of scientists who will tackle problems presented by the ocean as a whole, including the effects of man in the system.

CHAPTER 14

POLLUTION

B.H. Ketchum (*Group Leader*)
P. O'Kane (*Rapporteur*)

C. Morelli   E.D. Schneider
J.L. Hyacinthe   J. Veiga
S. Buckingham   C. Pissario
D.M. Di Toro   A. Rozeira
M. Bernhard   R.F. Henry
W. Bayens   E.D. Goldberg
I. Elskens

14.1. POLLUTANT CHARACTERISTICS

The pollutants of greatest concern in the marine environment are those which are toxic, which persist for long periods of time in the environment, and which reach the environment in large enough quantities to damage the marine ecosystem, challenge public health through exposure or ingestion of sea food or jeopardize amenities. Soluble persistent pollutants are called conservative since their distribution is governed by the same physical processes that control the circulation of the water. Modelling the distribution of such pollutants is within the present state of the art, wherever physical circulation models are available. Other types of pollutants are *degradable* and change with time as a result of biological, chemical or geological processes. For these, it is necessary to know both the characteristics of the circulation of water and the rate at which the processes which modify the pollutant character are taking place.

Table 14.1. lists various categories of pollutants with notes about the time and space scales and the present state of modelling applicability. The likely pollution problems that may arise in the next decade, based upon toxicity, annual production, and persistance in the environment of the materials involve:

(1) the transuranic elements, neptunium, plutonium, and curium, produced in nuclear reactors and leaking therefrom as well as from nuclear reprocessing plants, nuclear energy sources and during shipment;

(2) marine litter such as plastic ware, glassware, metal objects that are accumulating in surface waters, on beaches and on the bottom;

TABLE 14.1

Some pollutants with their modelling status

Pollutant category	Time scale (π sec)	Space scale	Modelling status
biodegradable wastes, enteric bacteria thermal	days—weeks (10^5)	local	satisfactory in some estuaries and bays
chlorinated hydrocarbons	decades (10^8)	global	exploratory global
heavy metals	millennia (10^{10})	local to global	exploratory some local
petroleum hydrocarbons	hours—years $(10^3 - 10^7)$	local to global	none
viruses	seconds—hours $(10^0 - 10^3)$	local?	none*
radioactivity	seconds—centuries $(10^0 - 10^9)$	local to global	exploratory local to global

* Some physical models can define maximum or worst conditions.

(3) synthetic organic chemicals such as low-molecular weight halogenated hydrocarbons;

(4) agricultural, industrial and hospital wastes which are discharged into patrimonial waters.

Models to predict accumulations of such substances in various components of the marine environment are urgently needed by management. The development of protocols to regulate the use of materials that can threaten the health of the oceans will depend upon such models.

14.2. SIZE AND SCOPE OF MODELS FOR POLLUTION PROBLEMS

A computational complexity

If one considers N, defined as the product of the number of state variables and the number of spatial compartments, then it is suggested that the range of simplicity vs complexity of models for solution of pollution problems spans the range of $N = 10 - 100$ for elementary models through $N = 100 - 1000$ for moderate complexity to $N = 1000 - 10,000$ for complex models.

As a rough guide to computational complexity one hour of central processing unit time on a modern large computer (e.g. CDC 6600, UNIVAC 1108) can execute on the order of 10^6 variable-compartment evaluations. Thus for a seasonal time scale (1 year) and an integration interval of 0.1 day a single execution for a complex model ($N = 5000$) can utilize approximately 20 h ($5000 \times 4000 \times 10^{-6}$) of central processing unit time. This appears to be a practical upper limit for the utilization of present-day computer time.

Economic considerations

The costs associated with water pollution control procedures, standards and facilities usually run in the millions of dollars. It is vital that the facility meets the water quality objective established in the estuary or coastal receiving waters, since the costs associated with under and over design can be quite large. The cost of modelling, as opposed to the costs associated with normal data collection and planning are usually a relatively small fraction of the total. Yet without a modelling framework and analysis, it is difficult to analyze the collected data and to forecast impact of the alternatives available to the decision makers. The management of the quality of estuarine and coastal waters requires estimates of probable impact of certain management alternatives. This is best supplied in quantitative terms by the use of a modelling analysis. Simple steady-state, one-dimensional, pollution, multi-constituent, engineering models take 1—2 man-years to complete. Second-generation, time-dependent, two-dimensional, multi-constituent, engineering models take 3—5 man-years to complete. Complex hydrodynamic modelling can significantly increase the required effort.

Marine toxicology

Acute lethal toxicity tests have been made on some chlorinated hydrocarbons, heavy metals, radioactivity and petroleum hydrocarbons on various life stages of marine life. Long-term chronic studies are lacking. The extrapolations from acute studies to the real world ecosystem is accomplished by the use of an "application factor" which is at this time only an estimate of the levels which will protect the system.

Most of the effects of pollutant data collected to date has thus been using lethal effect as a measure of stress. However, in the past few years researchers have been attempting to determine the sublethal effects of pollutant stress. In particular, behaviour, growth, predation patterns, reproduction, enzyme balances, histopathology, disease and physiologic functions are being studied. As yet little hard data have come from these studies, except that "adverse" effects on some species appear for pollutant concentrations an order of magnitude less than lethal levels. Virtually no data exist on the synergistic effects of several stresses on an organism or on a marine eco-system. Work is just starting on pollutant stresses on whole systems.

The weak link in pollution modelling is lack of knowledge about effects of a pollutant or combination of pollutants on the total ecosystem. Small per-turbations may cause major changes within the delicate balance of an eco-system. For instance, adding a chelating agent may change the metal balance in the water, perhaps affecting the species composition of the phytoplankton population. Variations in the base of the food web are amplified up to trophic levels, and the end products of changes have not been studied. The interaction of many variables create an ecosystem and any one of these variables may be critical to the function within a system. The importance of these forcing functions needs to be studied in many small systems before they can be combined in complex ecosystem models.

Coastal circulation concepts

Offshore currents, where the circulation is driven primarily by the wind, can be modelled in order to determine the fate of passive contaminants. Inshore currents, which are driven by the tides and friction, might be mod-elled by using offshore current models to provide appropriate boundary conditions. The formulation of appropriate boundary conditions and driving mechanisms is the area of greatest difficulty. The fate of passive contami-nants can be determined given a verified model of inshore currents.

In general, there are acceptable models for some estuaries which are useful for engineering purposes. These generally extend only to the BOD—DO prob-lem, and not to the biotic changes. Open-ocean models are satisfactory for gross mass balance estimates. The coastal water circulation is less predictable and there is no adequate conceptual foundation to permit modelling of the coastal zone circulation except where extensive, long-term, local observa-tional data are available.

14.3. CONCLUSION

Techniques for predicting the spatial distribution of pollutant input to the marine environment are quite well developed, but it is not possible to simu-late the effects of this pollutant distribution on marine organisms. The mod-elling techniques exist, but knowledge about the actual response of orga-nisms to pollutants is almost completely lacking.

CHAPTER 15

PRODUCTIVITY OF SEA WATERS

T.R. Parsons (*Group Leader*)
D. Menzel (*Rapporteur*)

G. Radach   C. Walters
N.R. Andersen   J. Steele
G. Pichot   D. Rodrigues
J.P. Mommaerts   R.C. Dugdale

15.1. INTRODUCTION

The need for models designed to define and predict the productivity of sea waters is evident. As aids in research, the combination of functional relationships between biological phenomena, and physical and chemical processes in the environment, allow the investigator to identify additional criteria to properly predict the productivity of a system and to evaluate its stability. Such models generally call for great sophistication and the input required relates directly to the demands of the researcher. On the other hand, relatively simple models can be derived from them to assist in decision-making processes. These are usually successful when applied to restricted areas for specific purposes. Such models can be built quickly and can be added to when other concerns become apparent (e.g. prediction of the consequences of nutrient loading in estuaries). However, the usefulness of management models for general application are limited by the quality and quantity of biological data imposed and the type of functional relationship used to describe the interaction between physical, chemical and biological components. In particular, the adaptation of the organisms themselves to changes in their environment is poorly understood and can greatly influence derived generalizations.

15.2. BIOLOGICAL ADAPTATIONS

Biological adaptations can be broadly divided into short-term *physiological responses* and long-term *selection*. Examples of the former are alteration to photosynthetic rate in response to changes in light intensity, or changes in the filtering rate of herbivores in response to changes in the composition or concentration of food. These types of responses occur within time scales varying from several minutes to days. Natural selection takes place over much longer periods of time (generations) and influences the life history characteristics and species composition of the population. Such changes may directly affect the structure of the food web although they need not necessarily affect total production.

Physiological adaptations

Data on the physiological responses of organisms to change in the physical and chemical environment have accumulated largely from single-species laboratory culture and have been validated in natural populations. As examples, functional relations have been established relating phytoplankton growth to nutrient concentration, the feeding rate and discrimination of cell size by zooplankton at different prey densities, species composition of the phytoplankton, and the temperature response of organisms, etc. However, the form of these functional relations now needs to be re-examined for such phenomena as threshold nutrient concentrations, the concentration of prey organisms in mixed populations, and the role of bacteria in nutrient regeneration and in establishing the base of neglected side food chains (e.g. bacteria—microzooplankton). Further evaluation of constants in the biological relations included in these equations, such as those relating to the uptake rate of different forms of nutrients (e.g. NH_4^+, NO_3^-, urea) by phytoplankton is needed. At higher trophic levels more information is required on the size selectivity and rate of consumption of prey by different predators, and on the early life history and changes in the spawning and migratory behaviour of juvenile and adult fish. Quantitative information is also required on benthic communities.

Selection adaptations

The adaptation of biological communities through selection is an aspect of oceanography which has been neglected. Even when dramatic changes have occurred in dominant species in given geographical areas (e.g. changes in herring and pilchard fisheries in the English Channel), the causes of these changes have not been understood. Considering the probable complexity of the underlying causes, which may couple biological, chemical and/or physical phenomena, modelling in such instances is usually after the fact. Such existing models are not predictive but explanatory and by necessity have been greatly oversimplified. This is so because present food chain models can indicate changes in energy flow resulting from environmental variations but cannot show equally or more important changes in species composition, particularly at higher trophic levels harvested by man. To provide proper perspective, food *web* models which separate trophic levels into two or more components are required. From simple mathematical (analytical) models of webs we know that gradual variations in coefficients can produce sudden changes in structure, leading to selection of different parts of the web as the dominant components. However, to produce simulation models of this type requires data on *differences* in growth and metabolic rates and selection in behaviour patterns within trophic levels rather than the *averages* which are used at present. Further, we know that such differences must be provided at each trophic level and that these probably become more important at higher levels. Some information is available for phytoplankton, less for copepods, but almost nothing for carnivores. Relatively simple food web simulations could provide guidance on the kind of experiments which might be most critical. At the same time one must realize that the testing of more complex models will require a comparable complexity requiring interdisciplinary field investigations.

15.3. CHEMICAL EFFECTS

Obviously, significant headway cannot be made in the matters identified above without considerable input from the chemical and physical disciplines. In the case of the former the control of production and behaviour of higher organisms is markedly influenced by alterations in the chemistry of sea water. The sophistication of chemical techniques is now adequate to describe simple nutrient kinetics etc. (e.g. N,P,Si), and major strides have been made in describing the control of production by these elements. Other microconstituents of sea water (trace metals and organics) are known as essential nutrients for plant growth, yet their availability may be directly governed by their inorganic and organic combinations, both ionic and non-ionic, none of which is understood, quantified or even identified. Additionally, animal behaviour, sex ratios, feeding rates, migrations, etc., may be governed by the presence or absence of trace amounts of organic compounds. Most of these have not been identified and modified behaviour is often attributed to them when other explanations are lacking. The study of the organic chemistry of sea water has been badly neglected, now being the most primitively developed of the chemical sciences of the ocean. Biologists can recognize these problems but there are few solutions to proper quantification of the variables. Frequently the most fruitful leads are those which arise from the interaction of physiologists, microbiologists, chemists and ecologists.

15.4. MODELS

Most models for examining the productivity of phytoplankton and grazers in the oceans are based on systems of differential equations for change in nutrient levels and biomass. Most consider at least one spatial dimension (depth or distance along a current). Few models have considered effects of the full three-dimensional hydrodynamic regime on production processes. Conceptually this extension would not be difficult, but we have been largely slowed by practical problems of computation speed and reasonable estimates of hydrodynamic velocity fields.

Hydrodynamic and biological modelling should proceed in parallel. Future research might well be centered on the introduction of more complex three-dimensional hydrodynamic regimes (full velocity fields, diffusion) into the basic production models, but there are other areas that should also receive attention. One of these areas is the possibility of using other mathematical techniques, such as mixed difference differential equations (model systems with both differential and discrete time equations) with different time steps for different components. This will improve computational efficiency and assist in representing some of the essentially discontinuous features of biological systems. Another important area where modelling work is lacking is the coupling between basic productivity and higher trophic levels, largely because of the biological complexities introduced by behavioural mechanisms in high animals (e.g. migration, prey selection).

Another major area for research is model validation. We need basic philosophical guidelines, for example, to compare model output to time series data. We know from the equifinality principle of general systems theory that many alternative (and even contradictory) models may produce the same time dynamics for some parameter values and initial conditions. We badly need procedures to help identify such pitfalls, and we should demand that any model be tested in a wide variety of situations. It is obvious that no single model can be developed to answer every question. We need to examine the effects of omitting various kinds of information in different models.

15.5. TRAINING AND FACILITIES

From the biological point of view it is evident that the training of plant and animal physiologists who can understand problems related to adaptation and its causes is badly needed. Such physiologists should be sensitive to the demands of modellers, but more importantly modellers should be cognizant of the realities of biology. This can best be accomplished by the cross training of, or cooperation between, biologists, chemists and physicists. For research guidance productivity models have frequently been useful only to those people who build them. It would be more efficient if oceanographers learned enough about modelling to take full advantage of research suggestions provided by mathematicians and engineers. Particularly, there is need for additional training courses and workshops on the techniques of model building specifically geared to the backgrounds and interests of oceanographers. Such training courses have been successful in several institutions.

Finally, a major need exists for large-scale compatible computer facilities in Western Europe committed to oceanographic research and management. A single installation of the most powerful system available, with a series of high-speed remote terminals, would provide facilities to a wide variety of people working both on regional models (e.g. estuarine pollution cases) and on general models of international interest.

CHAPTER 16

ESTUARIES

R. Wollast (*Group Leader*)
H.G. Gade (*Rapporteur*)

J.D. Burton   H. Postma
J.J. Leendertse   R.E. Ulanowicz
K. Mountford   W.M. Chiote Tavares
H.T. Odum   Isable Ambar

16.1. MODELLING IN ESTUARINE SYSTEMS

In the following an attempt is made to outline the specific problems of modelling of estuaries as characterized by the discharge of fresh water into a partially enclosed sea water body.

The hydrodynamical regime and exchange mechanisms encountered in estuaries lead to specific chemical, biological and geological processes requiring specially adapted models.

16.2. MODELLING OF PHYSICAL ESTUARINE PROCESSES

A wide variety of estuarine models has been developed; the more simplified ones, which are mathematically more tractable, involve averaging properties over space and time. These over-simplifications of the models may well violate the basic physics of the circulation processes, but still give reasonable representations of overall observed distributions and are used to compute the flushing time or the exchange ratio for the estuary.

More elaborate models based on the equations for conservation of mass and momentum and taking into account the geomorphology, the tidal flow and the river flow have further been developed to evaluate the physical processes within estuaries which produce the observed distributions.

Simulation methods

The one-dimensional approach is well developed within the accuracy of the parametric expression of mixing. Some question remains concerning conservation of mass in presently used computations. The two-dimensional approach is coming out of the experimental phase into operational computations.

Two methods of coupling the hydrodynamics with the transport processes can be used. These are the computation of the hydrodynamics simultaneously with the advective diffusion equation, both computed with the same timestep, and the computation of the advective diffusion equation with a much larger timestep than in the hydrodynamics equation. In the latter case serious questions concerning conservation of mass can be raised. Attempts to solve the baroclinic problem with numerical methods have not yet been developed into an operational method.

Three-dimensional computations are now being approached by finite differences over the vertical in addition to the horizontal finite-difference approximation and by analytical solutions of the vertical velocity field.

Presently no three-dimensional model of estuaries is operational, but further development using computers now being made can lead to operational models which can be used for scientific inquiry and solution of most engineering and estuarine management problems.

Analytical methods

Analytical methods are at present inadequate for the solution of engineering and management problems, but they contribute considerably to scientific inquiry.

Presently diffusive transports are introduced into the computational methods by parametric expressions derived from empirical relations. Further inquiry into the nature of these relations is urgently required. This inadequacy is particularly evident in deep estuaries where multi-layered velocity fields are prevailing.

16.3. MODELLING OF CHEMICAL ESTUARINE SYSTEMS

For modelling of chemical processes the important examples of state variables are salinity (which is used as an index of mixing), temperature, p^E, concentration of dissolved oxygen, pH, alkalinity, concentrations of organic matter, nutrients (P, N, Si), certain metals and particulate material. Other variables may become important in specific situations (pollution). For the correct expression of state variables, a knowledge of chemical speciation is needed. In practice there is usually a crude distinction of operationally defined fractions (e.g. dissolved, particulate, organic) but ideally the true chemical speciation should be known. This is a more complex task for estuarine waters than for oceanic and fresh waters because of the pronounced gradient of ionic strength.

For conservative (non-interactive) constituents their distribution is more readily modelled as it may be computed from the mixing properties provided by the physical model. For non-conservative (interactive) constituents there may be special boundary conditions to consider because of possible significant exchanges between the estuarine water and (a) the bottom sediments and interstitial waters and (b) the atmosphere. A chemical model will normally incorporate biological processes. We have little information on rates of uptake and release by organisms which is applicable to the highly variable conditions in estuaries. Similar considerations arise with the equally important task of incorporating interactions between dissolved and suspended particulate matter. These processes can often be described by parametric expressions without detailed knowledge of the individual processes concerned, but such knowledge may sometimes be needed to understand the mechanism. By comparing predictions based on various formulations of such parametric expressions with the real situation, greater insight may be gained into the factors controlling such biological and geochemical processes.

Sedimentation should be considered as an important removal process. Therefore, we recommend that models of sedimentary processes (flocculation, deposition, resuspension) be developed.

From the sedimentologist's standpoint a rough distinction can be made between movements of a coarse fraction (sand) and a fine fraction (silt). In the case of appreciable sand movements the morphology of the system is also changed and this in turn causes modification of the hydrodynamic system. No models for these interactions are available. In the fine fraction it would be of great importance to understand the dynamics of formation of the so-called turbidity maximum, including the processes of grain size selection.

16.4. ESTUARINE BIOLOGICAL MODELLING

The history of estuarine biological modelling is long and varied. Models of single organisms and processes have often been quite detailed aids to understanding. In keeping with the purpose of this conference, however, consideration will be limited to ecosystem models which can be combined with physical and chemical models to yield a total simulation of the estuary.

In what follows, the ecosystem models under discussion have time as the independent variable. This is not to imply that the spatial variation has been excluded. Rather the models are considered as distributed sources and sinks which may be inserted into the equation of continuity of species to yield the full temporal and spatial distributions.

Estuarine ecosystems models are generally of low resolution, usually employing combinations of the broad categories listed below as model compartments.

Biotic	Abiotic
phytoplankton	nutrients
zooplankton	temperature
nekton	salinity
microbes	turbidity
benthos	detritus

The combination of interactions among the above state variables will vary with the investigator and the estuary being modelled. The totality of the mathematical statements made about these interactions is small and tends to be constrained, perhaps needlessly, by historical modelling efforts.

Investigators are urged to be liberal in experimenting with the spectrum of little-used or novel functions as better descriptions of the mechanisms of interaction.

The values of the coefficients used in the mathematical statements are determined by: (1) independent laboratory measurement; (2) educated guesses; (3) regression to prototype data; (4) budget computations; and (5) combinations and/or iterations upon the above.

16.5. CONCLUSIONS

So far, most simplifications have been done by considering separately the chemical, biological, physical and geological subsystems. Also needed are models that are simplified in resolution without eliminating the main physical, chemical, geological and biological processes. The highest priority should be given to the development of such ecosystem models.

There exist very few truly verified estuarine models. This paucity is due largely to the costs involved in obtaining quality prototype data. Certainly prototype data should deal adequately with the time and length scales involved. At best, all data should also be taken simultaneously. Obtaining such copious data requires prodigious man-power, equipment, vessel use, analytical facility and computer software. Nevertheless, such coordinated efforts are essential if the estuary is even to be understood in the holistic sense.

CHAPTER 17

ENCLOSED SEAS

B.O. Jansson (*Group Leader*)
K.H. Mann (*Rapporteur*)

N.S. Heaps	E.J. Fee
C.H. Mortimer	Ph. Polk
T.S. Murty	Y.A. Adam

17.1. WHOLE-ECOSYSTEM MODELS

Enclosed seas lend themselves particularly well to the study of the whole unit as an ecosystem, for the following reasons:

(1) Boundary conditions are usually relatively well defined.

(2) Nutrient, salt and water budgets can often be framed with more precision than elsewhere.

(3) Small basins lend themselves to whole-system field experiments.

Moreover, from the practical viewpoint, enclosed seas often serve as waste sinks and give rise to serious management problems, such as conflict of interest between waste disposal and recreation or aquaculture.

17.2. HYDRODYNAMICS OF ENCLOSED SEAS

A fundamental component of a good model of an aquatic ecosystem is the hydrodynamic structure. We recommend that support be given to continued development and refinement of fundamental hydrodynamics. Particular attention must be paid to processes contributing to mixing — for example, turbulence, unstable shear flow, short internal waves — because mixing is usually the most important direct influence of hydrodynamics on the biological system.

The resolution of a model is governed by the scale of its grid or box, but processes on a smaller scale are essential to the model and their treatment should be improved. For example, turbulence, particularly in stratified bodies of water, is of fundamental importance in biological processes; it is recommended that more attention be paid to parameterization on scales smaller than the grid of the model.

In many cases two models are required:

(1) A multi-layered model for reproduction of transient phenomena over periods of 2—3 weeks at various times of the year. The purpose is to determine the conditions of flow, involving the parameterization of small-scale motions. There is lack of reliable knowledge of the values of empirical coefficients (diffusion, drag, etc.). There also appears to be no reliable method for the prediction of turbulence, other than criteria involving such coefficients as the Richardson number and empirical connections with current shear and with friction.

(2) The second type of model is one which simulates long-term changes in flow and elevation; for example, intermittent exchanges between the enclosed sea and the outside ocean. At present there appears to be no established modelling technique available to simulate such long-term changes. It is recommended that work be done to develop such a technique.

Outputs of the hydrodynamic models will be specification of flow patterns under varying conditions of forcing particularly by forces of meteorological origin. These flow fields will include circulation patterns, upwelling phenomena, long surface and internal waves, and the formation, movement and dissipation of ice. These outputs provide the basis for enhanced understanding of the physics of the system, and serve as the essential framework for an integrated model of an ecosystem.

17.3. BIOLOGY AND CHEMISTRY OF ENCLOSED SEAS

It is clear that many important chemical processes are inseparable from biological processes and must be treated with them. The working group had a division of opinion as to whether all relevant component processes can be sufficiently well defined on the basis of existing knowledge to give reliable prediction through dynamic modelling of whole ecosystems. The possibility of doing so should be advanced by the following means:

(1) Study and modelling of biological/chemical subsystems should be encouraged. Examples are: (a) in relation to plant nutrients: plant-nutrient dynamics, nutrient regeneration by consumers, regeneration of nutrients from sediments; (b) food transfer relationships, as influenced by particle size, dispersal and chemical composition; and (c) accumulations of toxic substances in food chains.

(2) Encouragement should be given to first steps in integrating subsystem models into combined dynamic simulations, such as hydrodynamics with nutrient cycling or primary production with secondary production. Particular attention should be paid to interactions between components in different subsystems and to appropriate meshing of time and space scales.

(3) At the same time, there should be further research into whole-ecosystem modelling, bearing in mind that polluted and otherwise perturbed ecosystems require special attention. Particular topics for research include changes in stability and radical changes in species composition.

In each of the above stages progress in modelling must proceed hand-in-hand with experiments and field verification, through collection of relevant physical, chemical and biological data observed as nearly simultaneously as possible. For many such field operations, large coordinated programmes will be required. The design of such programmes should depend on the results of preliminary modelling and the results of the programmes should be used to modify and improve the models.

17.4. ADVICE TO MANAGEMENT

Concerning the feasibility of producing reliable, predictive, total-ecosystem models at the present state of knowledge, it is essential that when advice is given to management authorities, a statement is made about the state of the art: the potentialities for prediction and the degree of confidence in any prediction. Particular situations respond best to particular types of model and, while in some cases full-scale simulation may be possible, in others statistical predictions based on previous measurements may still be the best guide to management.

17.5. INTERDISCIPLINARY COLLABORATION IN ECOSYSTEM MODELLING: TRAINING AND IMPLEMENTATION

Education

We recommend the holding of short courses and the production of texts designed to train physicists, chemists and biologists in each other's relevant concepts and terminology. It is envisaged that the text might have a section on geophysical fluid dynamics for non-specialists, and similar sections on aquatic chemistry, relevant aspects of geochemistry, aquatic biology and ecosystem function.

There would also be merit in short courses dealing with techniques in statistics such as time series analysis, multi-variate analysis, stochastic methods, etc. The availability of such courses in particular institutions should be made known among interested persons.

Planning and implementation

Planning and implementation of a collaborative programme should include the following stages:

(1) A review of existing data and an initial attempt at modelling leading to

(2) Optimization of field programmes with respect to improved data coverage, coordinated collections, standardized and automated methods.

(3) Experiments in laboratory and field to elucidate mechanisms not sufficiently understood.

(4) As a consequence and parallel development, successive improvements in modelling.

Equipment needed for such a programme should include ships and aircraft, instruments for automated data collection (e.g. moored buoys) and access to the largest available computers.

There is need to give particular attention to the development of automated methods of collecting biological data.

CHAPTER 18

SOME TECHNICAL AND MATHEMATICAL ASPECTS OF MODELLING
OF MARINE SYSTEMS

D. Garfinkel

Mathematical modelling involves the application of generally valid mathematical and computer techniques to specific subject matter with its own specific problems. Modelling of a specific subject area, such as marine systems, will thus have a different flavour and somewhat different techniques from other areas.

However, there are some common truths which apply generally:

Computer and mathematical techniques do not create information — they only transform it. The quality of the output will depend strongly on the quality of the input data (as the computer industry expresses it, "garbage in garbage out").

It is difficult or impossible to build a realistic model in any subject area without expert knowledge of that subject area, including expert intuition. However, models sometimes result in "counter-intuitive" behaviour, often in the form that a system goes in the right direction but overshoots or oscillates. This behaviour should be understood before it is acted on. The converse is also true: experimentalists should not edit or change experimental or observed data solely because it is intuitively unattractive.

All models are derived on the basis of a set of assumptions, which should be clearly stated, and are valid within a limited range of parameters. Extrapolation outside this range is always less reliable than interpolation within it, but may be necessary for management functions, where it may be necessary to predict how a system will respond to extreme or stress situations.

For research purposes one often learns more from transient situations than steady-state ones. However, management functions may often be possible with steady-state models. Automatic techniques of taking a specific subset of a research model to use for management functions may be possible.

Marine system will have the following specific problems:

We have to consider problems with physical, chemical and biological components. For the first two, the models are relatively determinate and robust (or they can be treated in this way as a result of available techniques such as parameterization), and the data relatively accurate, as compared to the biological elements. These appear to be highly stochastic (often because we do not know what is determining them in detail, or exactly what the real decision points are), not robust (i.e., they are sensitive to the exact numbers), and sometimes hard to define. Although any water molecule has the same properties as any other, one mussel may not be a substitute for another, owing to differences in size, etc. Apparent stochastic behaviour is exemplified by the classic work of Park with flour beetles, and the work of Paine and of Dayton on intertidal communities, where one sometimes sees adjacent identical tidal pools with vastly different flora, and where there is apparent non-reproducibility of experiments, owing to such random disturbances as cold weather, battering by logs, or the entry of a new species which strongly affects the behaviour of others. It has been suggested that heterogeneity, both within species and in the local environment, may stabilize systems, in that some important members of a system may be free to migrate away from unsatisfactory environments.

Marine systems are spatially inhomogeneous, and this is often dealt with by dividing a system into many box models which are usually concerned with the same entities. This greatly increases the computer time required.

Management models are required to be reliable but need not be (and probably cannot be) very accurate (certainly less accurate than econometric models are apparently expected to be).

We are dealing with systems that are inherently complex. As a result, fitting of models to data by the standard optimization techniques will be extremely expensive, if it is possible. Also, data will generally be scarce compared to the need for it, and it would be desirable to have improvements in the statistical techniques for scanty and incomplete data.

Some recommendations for dealing with these problems are possible:

(1) It will be necessary to make provision for the stochastic element introduced by the biology. This may be done by modification of either existing techniques for discrete-event or continuous simulation (solution of differential equations).

(a) Observations such as those of Park and Paine could be modelled by extension of existing discrete-event simulation languages, or perhaps even effective use of the "richer" existing languages. Specific abilities required here are: (i) use of space as an important variable, with competition for space as an important part of the model; (ii) ability to effectively keep track of individuals in moderate numbers, including change in properties with time (e.g. growth); and (iii) ability to keep track of and respond to changes in the environment (by modifying model constants).

(b) It is possible to introduce random events, such as single events of catastrophic scale, or continuous random events, in continuous models and see how these are affected. Some of this could be done by adding to continuous models methods currently used for discrete-event models. Some additional techniques may have to be developed; others may be adapted from meteorology.

(2) Fitting of complex models may be made much less costly by development of methods for breaking them into pieces (e.g. individual species or groups of species such as "zooplankton"), optimizing the pieces, and putting them back together into a complete model. We have had encouraging success with this tactic in modelling complex biochemical systems (which are technically similar to ecological systems).

(3) The expense of dealing with spatial inhomogeneity could be reduced by careful attention to whether one should use partial derivatives in preference to box models or before using box models, how fine a subdivision into box models is needed, whether they should all be the same size (in particular it may be possible to use different sizes for different variables), whether a given box need be the same size at all times and whether the computation must include everything for all boxes (this could perhaps be varied with time in the same manner that integration step-length is now varied). Work on more efficient use of such techniques is now under way. This may require a moderate application of artificial-intelligent techniques.

(a) Since systems of box models will have the same biological entities in several or many boxes, calculation of their behaviour may be economized by representing each entity by a set of coding into which one can enter the different conditions rather than solving a different set of differential equations in each box.

(4) Probably a considerable cost-saving can be effected by adapting programs to the computational properties of these systems rather than using general-purpose languages. Savings may also be effected by making it easier for personnel involved to use the computer effectively — it is the total cost of a project that counts.

(5) It may make it materially easier to work with these systems if they can be described in the user's terms with the computer handling their conversion into mathematical form (i.e., one could write a problem-oriented language, although this would be complicated by the presence of physical, chemical and biological elements). These systems may get complex enough to make it hard for the user to keep track of what is going on otherwise. This has been necessary for complex biochemical systems. However, it may be unprofitable to do this before the users understand just what they want to do, especially if they are still working with simple models.

(6) Some specific ecological modelling problems that might deserve attention include:

(a) Formulation of a method of deriving relative rate constants from the competitive behaviour of species.

(b) Data on biological populations usually gives population as a function of time. It has been suggested that the derivatives of the populations may be important. Just what should we be looking at here, and how?

(c) Especially for management purposes, can one compute some kind of measure of stability or instability? (It is recognized that this is still in controversy at the research level.)

(7) Simulation of large systems (primarily if it involves solving differential equations) may be greatly helped by the ability to interact with a large computer while the behaviour of the model is being calculated. This is more readily available in North America than in Europe. The difference is less in lack of appropriate computers than in telephone systems unaccustomed to long-distance data processing and, perhaps as a result, computers unaccustomed to effectively working in time-sharing. Remedying these deficiencies would be worthwhile.

INDEX